P9-EEP-763

THE INTERNATIONAL LIBRARY OF
SYSTEMS THEORY AND PHILOSOPHY
Edited by Ervin Laszlo

THE SYSTEMS VIEW OF THE WORLD
*The Natural Philosophy of the New
Developments in the Sciences*
by Ervin Laszlo

GENERAL SYSTEM THEORY
Foundations—Development—Applications
by Ludwig von Bertalanffy

ROBOTS, MEN AND MINDS
Psychology in the Modern World
by Ludwig von Bertalanffy

THE RELEVANCE OF GENERAL SYSTEMS THEORY
*Papers Presented to Ludwig von Bertalanffy
on His Seventieth Birthday*
Edited by Ervin Laszlo

HIERARCHY THEORY
The Challenge of Complex Systems
Edited by H. H. Pattee

THE WORLD SYSTEM
Models, Norms, Applications
Edited by Ervin Laszlo

A GENERAL SYSTEMS PHILOSOPHY
FOR THE SOCIAL AND BEHAVIORAL SCIENCES
by John W. Sutherland

DESIGN FOR EVOLUTION
*Self-Organization and Planning
in the Life of Human Systems*
by Erich Jantsch

Further volumes in preparation

DESIGN
FOR
EVOLUTION

Photograph by Angela Maria Longo

"Rising to the Sixth Chakra"

DESIGN
FOR
EVOLUTION

*Self-Organization
and Planning in the Life of
Human Systems*

ERICH JANTSCH

GEORGE BRAZILLER

New York

FOR ANNE

Le vent se lève, il faut tenter de vivre
(The wind rises, one must try to live)

Paul Valéry

CONTENTS

PART I
The Challenge of a Dynamic World

PART II
Design: The Human Way of Relating to the World

PART III
Consciousness Evolving

PART IV
The Appreciated World Evolving

PART V
Reality Evolving

INTRODUCTION AND SUMMARY

This book is an exploration into the acquisition and use of knowledge for human purposes. It elaborates on the theme—not novel to our critical age—that rational knowledge cast into the form of internally consistent, logically constructed and closed models of science constitutes a useful, but by no means sufficient, tool for dealing with matters of human design for a human world. In such matters, the *"know-how"* of ordering and implementing well-perceived, goal-oriented action, the very domain in which the objectivating models of science are helpful, constitutes the lowest level in a hierarchy of knowledge for human purpose.

As Norbert Wiener emphasized more than two decades ago, the *"know-what,"* the setting of goals in a systemic context, is infinitely more important. But it would be a blind art if it could not receive any guidance from still a higher level. This is the level of the *"know-where-to,"* at which we strive for an overall sense of movement and direction, of the dynamics of our individual lives as well as the lives of human systems of all scopes and sizes, and of humanity in general. It is also the level of *policy design*—of the design of enduring themes and regulating principles underlying these lives. To what extent and by what means we shall be capable of shaping our own future—that is, of regulating the dynamics of human systems—will be decided by the understanding we can bring to this level.

With Ervin Laszlo we may say that having addressed ourselves to the understanding and mastering of *change*, and subsequently to the understanding of order of change, or *process*, what we now need is an understanding of order of process (or order of order of change) —in other words, an understanding of evolution.

It is this highest level of the "know-where-to," of the design and regulation of dynamically evolving systems and processes, to which this book is addressed. Thus, it is hoped it will enrich the perspective and vision of all people responsible for and interested in the design, regulation, and restructuring of the lives of human systems at all levels: managers in government, social institutions, and business; academics developing interdisciplinary approaches to human-, society-, and mankind-oriented studies; forecasters, planners, and designers; philosophers and philosopher-kings; artists; individuals engaged in their own spiritual growth process; and, above all, young people in search of a higher sense of purpose than the one provided for by current cultural stereotypes. It is, in fact, my interaction with young people in and outside universities that motivated me to develop a fresh and unorthodox approach and gave me the courage to explore the subject of human design in accordance with urgent needs, which means to a fuller extent than would be possible on the crumbling grounds of contemporary established science.

But it also became increasingly clear to me in my frequent interactions with leaders in government and business that the secret of the successful manager can never be fully, or even predominantly, explained by mere efficiency in information processing of a conventional kind. What we are used to call "fingertip feeling," or intuition, or simply luck depends in reality on the capability to "tune in" to an overall movement, to influence and subtly regulate powers beyond those vested in the human intellect, and even in personal charisma. The successful manager inexplicably is "with the stream," whereas the unsuccessful one tries in vain to develop a rationale which would be capable of telling him about the movement of the stream. It is this most important, but almost totally neglected—because nonrational—aspect of management science to which this book hopefully will make a contribution.

Thus, the development of a sense of purpose and direction is my main theme. Anachronistically comes the advice which a British prime minister lately saw fit to give his country in benign disinterest: "If people want a sense of purpose, they should get it from their archbishops." We cannot leave this question to a static authority any longer; in a time of major transitions, such as ours, we have to make it our foremost concern and all-important personal task. Even the more innovative literature on management, planning, and systems design—and my own writings up to now are no exception—restricts itself to the structuring of a rational framework for information processing and decisionmaking. Above the clouds usually hovers a notion which is taken to represent the nonrational, or even the irrational side of man, the inexplicable and uninfluencable, the abysmal, denoted by the word "values."

The whole of management science bases its legitimacy on a kind of Indian fakir rope trick—standing on the allegedly firm grounds of "facts," data, and rational "know-how," it throws ropes into the air and makes believe that we can climb them up to the clouds which is where management science starts to lose interest. The assumption that the rope trick is real, constitutes the metaphysical basis of management science and all forms of management—and a pretty shaky basis it is as it turns out in our time of major transitions. But what could we do to realistically anchor the rope in the realm beyond the rational, in values and human archetypes, in a purpose and meaning elusive of rational analysis? Contemporary science does not venture there; it even busies itself with building walls to block off those areas from which it is ultimately nourished.

I believe that there exists an urgent need to explore the dynamics of those areas "beyond the clouds," where human cultures are rooted and from where *cultural change* emanates as the powerful guiding factor of social and individual change. In attempting this task, I took equally seriously all modes of holistic insight: paradigms of physical and social science, artistic expression, mystical insight, and psychic revelatory processes. This does not imply that holistic forms of knowledge will not eventually, to a much larger extent than at present, be cast into scientific terms. But time is pressing—we have to use knowledge in whatever form it is coming to us. Therefore, and without apologizing, I frequently continue my text by quoting from poetical expressions where scientific formulation runs into limitations. In doing so, I was able to draw on an ageless wisdom that may guide us in our groping for deeper understanding, while also using the terms we have developed in a scientific/technological culture.

Thus, I approach the task of conveying the fuzzy outlines of an emerging holistic and dynamic vision of the human world by painting *al fresco* rather than by developing details for which the arguments are not yet all that clear in scientific terms. My principal aim is to get a fair grasp of a new paradigm, a new regime of thought expressing the life of human systems—in other words, a mutation which transcends the possibilities of any compelling logical and closed argument developed by the rules of an established regime. I hope that in this way I also make available to the reader a richer body of knowledge with parts of which he may get into resonance, and a better readable and more interesting text.

Evolution and design, the course of nature and man's intervention in it, are notions that seem to clash in the dualistic view taken by Western thought. Human action is usually set off against all other movement in the universe. Or if it is recognized as an expression of life at large, the latter is viewed by conventional Western science as a specific, rare, and, in the end, futile process which pushes uphill against the broad stream washing downhill toward increasing randomness and entropy. In such a dualistic view, human life finds its meaning in the margin left between the attitudes of Promethean rebellion and devout fatalism.

Recent breakthroughs in physical science, in particular in the field of nonequilibrium thermodynamics, point the way toward overcoming the duality of older models. I am referring particularly to the discovery, by Prigogine and coworkers, of the principle of "order through fluctuation" which seems to govern the evolution of physical as well as of biological systems. In this book, I am trying to extend the application of the same principle to human systems, in particular social and cultural systems, but also to knowledge systems, to the development of human consciousness and to the human design process. Behind these intuitively guided hypotheses, which are open to scientific investigation, lies the great hope of laying a new and modern foundation for a profound truth which man has known, forgotten, refound, reduced, and expanded over many millennia: that *the evolution of mankind forms a meaningful and integral part of a universal evolution—that mankind is an agent of this universal evolution, and even an important one.*

If "order through fluctuation," indeed, turns out to be a basic kind of mechanism for the unfolding of evolutionary processes in all domains, a unified view of evolution becomes a distinct possibility. It will make the old dream of general system theory come true in a new light. The unifying principle will be found in the dynamic conditions of nonequilibrium systems and the ensurance of continuous metabolizing, entropy-producing activity and energy exchange with the environment. It will no more be sought in the static or steady-state conditions of equilibrium. Open, or partially open systems in all domains—from atoms to galaxies, bio- to social organisms, human consciousness to cultures and mind at large—will then be carriers of an overall evolution which ensures that *life continues*, that a nonequilibrium world evolves to ever newer dynamic regimes of complexity. Life itself takes on a new and very broad connotation in this light, far beyond the narrow notion of organic life. Physical systems, human systems, mind systems, all have life in this broad sense. They do not behave randomly but mutate toward new dynamic regimes whenever they become stifled by the debris of past entropy production. The implication that physical energy in nonequilibrium systems is not monotonously running down to a state of maximum entropy (as the Second Law of Thermodynamics states for equilibrium systems) but in mutations always finds renewed possibilities for entropy production, points to prospects for a future unified concept of energy and its transformations. It becomes superfluous to postulate vital or psychic forms of energy which fight a losing battle against the degradation of physical energy. In an emerging unified concept, the same energy carries the evolution in all domains—whether it appears in the form of physical, vital, psychic, creative, or motivational energy, or in the form of matter, information, ideas, consciousness, complexity, or mind.

No area of inquiry will benefit more from such a new theory of evolving systems than social science with its currently dominant mechanistic and behavioral concepts. It will make it possible to deal in meaningful terms with the life of human systems, including the birth and evolution of cultures with their powerful influence on all human affairs beyond any rational grasp. It will also make it possible to deal with social as well as individual creativity—with *human design* in the broadest sense of the word.

An evolutionary perspective elucidates the continuous self-organization and self-realization of systems through processes which are in turn self-realizing and self-balancing. Internal and external factors of selection appear here as aspects of the same process, viewed from adjacent hierarchial levels; this complementarity generates hierarchical order in an evolutionary perspective. Again, these notions apply to all domains, but they are most richly orchestrated in human systems with their unique capability of introducing fluctuations and new energy in the form of ideas, expectations, models, myths and plans which surface within these systems. Quite generally, an evolutionary perspective emphasizes *process* over structure, the exchange of energy over its containment, flexibility and change over stability. It goes even further—it is interested in the *order of process*. Structure then is an incidental product of interacting processes, no more solid than the grin of the Cheshire cat. Its physical image is that of a standing-wave pattern as it may be observed, for example, in the confluence of two rivers; when the flows change, the standing-wave pattern changes, too. It is the processes which determine structure, not the other way around. And it is the ordered sequence of four-dimensional space/time events which determines the course of a process.

The book is divided into five parts.

Part I on *"The Challenge of a Dynamic World"* exposes some of the basic material to be dealt with in the book. In Chapter 1, I scan possibilities for the setting of policies, or rules for regulation, and conclude that the emerging task is not to select and fix yet another rigid policy for our time, but to understand that regulation in times of significant and rapid change depends on utter flexibility even in the principles applied to regulation. In fact, I go even further in my conclusions by developing in Chapter 2 the thought that not only can there be no one-sided resolution of the tension between opposites (as dialectical thinking has long recognized), but that this tension constitutes the very energizing factor keeping the human system in continuous motion.

A dynamic world view, as opposed to the static view engrained in Western science, focuses on processes and their interaction, processes of a self-realizing and self-balancing type which unfold in continuous feedback linkage to their own origins. Such processes are at the core of evolutionary world views which I survey in Chapter 3. In such views, it is the event unfolding in space and time which is set absolute; creation is continuously creating itself. I point to new insights into the mechanisms of physical and biological evolution, in particular to the above-mentioned principle of "order through fluctuation." The latter forms also the basis for my attempts, in Chapter 4, to link the idea of evolution to recent notions about the self-organization of human systems. The crucial question of communication mechanisms at work within human systems seems to lead to the recognition of resonance as perhaps the most important basic principle involved.

In *Part II*, *"Design: The Human Way of Relating to the World,"* I develop certain basic notions underlying individual, social, and cultural creativity, the sources from which the life of human systems feeds. Chapter 5 introduces as one of the basic concepts to be used throughout the book the triplicate structure of levels of perception, or inquiry—rational (separation between subject and object), mythological (feedback relation between subject and object), and evolutionary level (union of subject and object). Complementary views of reality result which permit the elevation of the reductionist view (quantity) obtained at the rational level toward a holistic (quality) and ultimately toward a nondualistic, four-dimensional space/time view (process).

In Chapters 6 and 7, I develop my model of the basic human design process which I graphically depict as the interaction of processes in a toroidal model. This model is composed of elements that are arranged in two planes as ternary systems. One of them links physical, social, and spiritual domains, the other consciousness, the "appreciated world," and reality in each of the above domains. The feedback interaction between the elements of such a ternary system, the possibility to conceive, nurse, and test alternative ideas in a malleable "appreciated world" before bringing them to bear on reality, makes creative design possible. Action is based on a ternary system, where mere response or behavior plays in a binary system between consciousness and reality. The rest of the book is structured in accordance with the ternary system of consciousness, "appreciated world," and reality.

Part III, "Consciousness Evolving," deals with the feedback processes shaping human consciousness. In Chapter 8, I trace some of these processes, forming an "outer way," back to notions developed by modern psychology. In particular, I try to show how the processes playing between consciousness and reality realize innate norms which become explicit at the three levels of perception/inquiry. Chapter 9, dealing with the approaches and techniques of an "inner way" to the direct and holistic experience of evolutionary movement and direction, introduces notions which are beyond the grasp of conventional science. The possibility of such a direct experience is based on the intuitive knowledge, or assumption, of a profound correspondence between universal and human dynamic order, in the spiritual as well as in the social and the physical domains—"as above, so below; as below, so above" in the succinct formulation of ancient Hermetic philosophy.

In Aldous Huxley's terms we are all Mind-at-Large, sharing in the same overall and universal evolving system of processes. I present a systematic overview of some sixty approaches which I structure in accordance with the four classic Hindu *sadhanas* (ways to enlightenment) and the three levels of perception/inquiry. I hope that even the incredulous reader will find it useful to have at hand such a new systematization of approaches on which an increasing number of people around the world, especially among the young, pin their hopes for overcoming the encrustations brought into our world by the exclusive claims of rational science and technology. In Chapter 10, I discuss the balancing of consciousness through apperception between the processes on both sides, reality and the appreciated world. The emerging structure of total experience—what we are, feel, perceive, know, want, conceive, and can do—may be represented in pairs of opposites which energize the evolution of consciousness and keep the value system alive. In fact, the resulting system of basic attitudes may also be viewed as a basic system of values that is continuously balancing itself in the interplay of tensions between opposites containing each other.

Part IV, "The Appreciated World Evolving," deals with some of the major processes by which we extend our basic experiences into the future—by which we plan. Chapter 11 focuses on the crucial feedback process leading to the formation of subjective models which then return in the disguise of objective myths. Here lies one of the most dangerous potential short-circuits in the design process which is capable of generating and freezing the most insensitive kinds of relations within our world. In Chapter 12, I review the basic planning processes in the light of a moral systems approach—in other words, under the postulate of an ethics of whole systems. If this postulate is still somewhat elusive as far as the evolutionary level of inquiry is concerned, a systems approach of this kind already gives a methodological framework to elevate planning from the merely rational to the mythological level of inquiry, the level where we live in most of our relations and where we experience the quality of life. The same elevation to the mythological level is aimed at by interdisciplinary inquiry and the broader inter- and transexperiential forms of inquiry to which Chapter 13 is devoted. Here, some basic conclusions are also drawn concerning the outline of a future approach to education for design.

Part V, finally, *"Reality Evolving,"* elaborates on some of the implications the new paradigms seem to hold for human systems. In Chapter 14, I view the organization of the human world as passing through successive "waves" of dominant forms of organization— from ecological over social and cultural to psychic organization— which also express the elevation toward human systems of widening scope, from individual systems to mankind at large and beyond. Each wave passes from a subconscious coordinative (internalized) phase through an increasingly conscious competitive (Darwinian, externalized) phase to a fully conscious coordinative phase—design. The interplay between internal and external factors of evolution seems to be at the core of a hierarchy theory of human systems.

At present, social organization is moving more deeply into the design phase and is approaching world unity, whereas cultural organization enters the competitive phase, and psychic organization is still largely subconscious. The elevation of the organization of the human world toward new dynamic regimes is in itself a striking example of the principle of "order through fluctuation," governing the macroaspects of the evolution of mankind. In Chapter 15, I try to discuss some of the implications for the phase the Western world is actually passing through, in particular the requirements for more flexible institutional role playing. I close with an outlook on the forthcoming phase of cultural design, and the emergence of new forms of human relations in the sign of universal love.

In the *Epilogue*, I identify planning and love as the two essential and complementary aspects of human design. Where planning, the masculine element, aims at stabilization which in turn makes it possible to act out power or focused energy, love, the feminine element, introduces the instabilities which elevate the plane of human action to ever new dynamic regimes, thereby ensuring the continuously renewed conditions for human creativity, for the life of human systems. Thus the design process itself seems to work by the ubiquitous basic principle of "order through fluctuation," like all aspects of evolution. Human design for evolution is not only a distinct possibility—it is also the foremost element of hope in times already marked by increasing fluctuations and turbulence.

I make no attempt to place the argument of this book into the full context of historic and contemporary thought with which it may link up in one or the other way. I simply quote those sources that I used directly. But I wish to acknowledge here some of the very profound and stimulating influences I experienced in writing this book. I found myself particularly open to these influences since they were not of an impersonal kind, but emanated from discussions and exchanges of view with personal friends whose work and search I deeply admire: Professor Ilya Prigogine of the Free University of Brussels whose guest lecture at my Berkeley seminar marked a turning point in my thinking and whose insight into the evolutionary

principle of "order through fluctuation" I adopted as the basic
paradigm for my inquiry into the evolution of human systems—also,
I owe many of the examples in this book to Professor Prigogine; Sir
Geoffrey Vickers, Goring-on-Thames, whose inspired and lucid
ideas on policy design, appreciative and value systems, and the roles
of institutions provided me with the best of all possible rational
platforms from which I could dare to venture out further; Professor
C. West Churchman of the University of California at Berkeley
whose untiring and warmly motivated search for human relevance
in planning, and in particular his philosophy of a moral systems approach, led me through many intellectual mutations onward to those
questions which I now finally attempt to take up; and Anne Parks,
Berkeley and London, to whom this book is dedicated and who
lovingly guided me to those areas of spiritual search and exercise
which form the nourishing grounds for an "inner way" to a direct
and holistic grasp of evolution and mankind's role in it. I used these
influences not in an eclectic spirit but let each of them guide me
from different angles to the heart of my subject.

I wish to acknowledge also the benefit of more specific stimulating
and clarifying discussions I had with a great number of friends
interested in particular aspects of my general argument, especially
with Professor Paul Feyerabend and Terence McKenna, both Uni-
versity of California, Berkeley; Professor Donald Michael of the
University of Michigan, Ann Arbor; Professor Ervin Laszlo of the
State University of New York, Geneseo; Milton Marney of George
Washington University, Washington, D.C.; Professor Hartmut von
Hentig, University of Bielefeld; Hugo Kückelhaus, architect in
Soest, Sissie Princess Bentheim-Tecklenburg, Rheda, Jens Tranum,
Lyngby, and many others. But above all, I want to thank the partici-
pants in the more recent seminars I led at the University of California,
Berkeley; at the University of Bielefeld, Germany; and at the Tech-
nical University of Denmark. It is through these interactions that
my thoughts took the shape found in this book.

I wish to thank also my dear friend Angela Maria Longo, doctoral
student at the University of California, Berkeley, for letting me
use the photograph for the frontispiece. Particular and deeply felt
thanks are due to my faithful advisor, language-editor and general
commentator, Gen Tsaconas of the University of California, Berkeley.

During my work on the manuscript I enjoyed successively the
hospitality of the Rockefeller Foundation at the magnificent Villa

Serbelloni in Bellagio (Italy); of the newly established Center for Interdisciplinary Studies at the University of Bielefeld, whose first resident I became; and of the Danish National Bank in the beautiful old house it maintains for foreign guests in the most picturesque part of Copenhagen.

ERICH JANTSCH

Berkeley/Copenhagen/London

ACKNOWLEDGMENTS

Chapter 4 includes extracts from a paper "Forecasting and Systems Approach: A Frame of Reference," published in *Management Science*, Vol. 19, No. 12 (1973). Chapter 12 uses large extracts from a paper, "Enterprise and Environment," presented at the Third International Management Symposium, St. Gall (Switzerland), May 1972; it has been published in German translation in the symposium proceedings, Benedict Hentsch and Fremund Malik (eds.), *System-orientiertes Management* (Haupt, Berne and Stuttgart, 1973), and in the English original under the same title, "Enterprise and Environment," in *Industrial Marketing Management*, Vol. 2 (1973). Chapter 13 is partly based and elaborates further on a paper, "Education for Design," presented to the symposium "The Universitas Project: Institutions for a Posttechnological Society" at the Museum of Modern Art, New York, in January 1972 and scheduled for publication in the forthcoming proceedings of the symposium, edited by Emilio Ambasz; smaller extracts from the original paper have also been used in Chapters 5 and 10; extracts from the original paper also appeared under the same title, "Education for Design," in *Futures*, Vol. 4, No. 3 (1972). Chapter 5, under the title "Transexperiential Inquiry," formed the basis for a prepublication presentation to the Second International Conference on the Unity of the Sciences, Tokyo, November 1973.

Large parts of Chapter 14, and additions from Chapter 15, have been used as the basis for a prepublication article, "The Organization of the Human World," in *Futures*, Vol. 6, No. 1 (1974).

The quotations from Chuang-Tzu in Chapter 2 (from the poem "In my End is my Beginning") and in Chapter 15 and the Epilogue (from the poem "Great and Small") are from the book: Thomas Merton (ed.), *The Way of Chuang-Tzu* (New York: New Directions, 1969). They are reprinted here with the permission of New Directions.

The Zen parables in Chapter 5 (from "Not the Wind, not the Flag") and Chapter 7 (from "Everyday Life is the Path") are from the book: Paul Reps (ed.), *Zen Flesh, Zen Bones: A Collection of Zen and Pre-Zen Writings* (Garden City, N.Y.: Anchor Books, Doubleday, no year; originally published in 1934). They are reprinted here by permission of Doubleday.

The extract in Chapter 5, from Act III of Witold Gombrowicz, *The Marriage* (New York: Grove Press, 1969), is reprinted here by permission of Grove Press.

The Dostoyevsky quotation in Chapter 1 is from: Fyodor M. Dostoyevsky, *The Brothers Karamazov*, translated by Constance Garrett (Chicago: Encyclopaedia Britannica, 1952). It is reprinted here by permission of Encyclopaedia Britannica.

The Saint Exupéry quotation which appears as motto for Part III, is from: Antoine de Saint Exupéry, *The Little Prince*, translated from the French by Katherine Woods (New York: Harcourt Brace Jovanovich, Harbrace Paperbound Library, 1971).

The Yeats quotations which appear as mottos for Part I (from the poem "Among School Children") and for Part III (from the poem "A Prayer for Old Age") are from the book, *The Collected Poems of W. B. Yeats* (New York: Macmillan, 1955).

The T.S. Eliot quotation which appears as motto for Part IV is from his play *Murder in the Cathedral*.

The Claudel quotation which appears as motto for the Epilogue is the motto for his play *Le Soulier de Satin*.

The photograph for the frontispiece is by Angela Maria Longo who calls it "Rising to the Sixth Chakra."

The painting "Cyphoma" which appears in the cover design, is by M. Purdy; it is used here by permission of the owner.

PART I

The Challenge of a Dynamic World

O body swayed to music, O brightening glance,
How can we know the dancer from the dance?
William Butler Yeats

IN a dynamic world of processes that include our lives and those of the systems in which and through which we live, where can order be found and enhanced—how can regulation be conceived and enacted? Is it the role of human design to fix a policy, a set of regulative principles, for the life of each human system? An evolutionary perspective, going beyond process, looks at the evolving *order of process* and tries to find there a sense of direction. The emerging picture is that of a nonequilibrium world in which physical, biological, social, and spiritual (cultural) systems all mutate toward new dynamic regimes in order to maintain their capability for energy exchange—in order to keep alive in a broad sense. A plausible basic mechanism, a new paradigm for a general system theory, is "order through fluctuation."

(1)

REGULATION,
THE HUMAN PREDICAMENT

Human life is movement. It is movement not by and for itself, but within a dynamic world, within movements of a higher order. On the one hand, these higher-order movements constitute the life of human systems—of relations, organizations, communities, institutions, nations, and cultures, of the entire human world. For human systems, no less than individuals, have a life of their own, a life which is characterized and enhanced by human qualities and capabilities. In our Western world, we have almost come to forget this since we started to draw organismic and mechanistic analogies for the interpretation of human systems. But on the other hand, human life is also embedded in the higher-order movement of nature, of an all-pervading flow of life and evolution which generates and energizes the lives of human beings and, through them, the lives of human systems. This we have forgotten to such an extent that, in order to remember and comprehend our own nature, we search in the myths of old and extinct cultures and in the depths of the subconscious—to emerge with but the vague contours of a remote reflection, an elusive memory from earlier phases of mankind's psychosocial evolution.

My preliminary image of human life, the life of an individual as well as the life of a system of human beings, is that of steering a viable course in the flow of a powerful stream which at times flows smoothly and majestically through wide plains, at times turns into wild waters rushing through narrow gorges and around sharp bends —reassuring and deceptive, supportive and endangering at the same time. It is the old Chinese image of *tao*, the right way, between the two opposing forces of yin and yang—the feminine and the masculine, the passive and the active, dark winter and bright summer. Tao was seen as an evolutionary principle which acts in the world as it creates itself, but it was considered man's task to shape his own fate by complying with tao and keeping the proper balance between yin and yang. In Chapter 5, I shall elaborate on this image and identify different aspects of it which correspond to different basic modes of inquiry, all of which are to be engaged in human ways of relating to the world.

Life is a gift and a predicament. It is the same gift for all creatures alive, but the particular human predicament differs in an important way from the general predicament inherent in life within the order of a natural ecosystem. The human predicament is regulation—not only living in accordance with, but *designing* forms of regulation which will ensure the steering of a viable course. There are many forces acting in the stream, pulling and pushing from all sides, sometimes mutually enhancing or canceling out, but more often conflicting with each other in complex ways. In the last analysis, these are forces acting within us—the conflict is within ourselves. But to us they appear as forces acting upon our lives from the outside, manifesting themselves not just in one vector, pointing in one direction at a time, but in a multitude of qualitatively different vectors—or, rather, in a system of relations which affect each other and whose interplay is to be regulated in many dimensions.

This, to me and others (Churchman, 1968 a; Vickers, 1970) constitutes the notion of policy: *A policy is a set of principles laid out for the purpose of regulating simultaneously and in a viable mode a multitude of interacting relationships pertaining to many qualities and dimensions of human life—in short, a theme underlying a life.* I believe it is not unduly restrictive to say here, "a viable mode" which excludes nihilistic attitudes toward life from our discussion; but there are, of course, policies that are at variance with the conditions of steering in the stream, that ultimately turn against life if kept rigid.

Vickers (1973) has pointed to the specific human character of policy seen in this light. Whereas an animal would attempt to deal with various relationships in sequential order, one after the other, humans can deal simultaneously with many relationships. For Vickers, the demands arising from multiple membership in many institutions have become the main characteristics of our complex times, thereby enhancing the need for regulation. The notion of multiple membership is indeed a very useful one for discussing dynamic human systems of a higher order, and I shall occasionally evoke it in this book.

There is no predictable security in movement, nor can perfection ever be reached and maintained. These are notions pertaining to a static, or steady-state, view of the world. Yet history is strewn with the consequences of man's longing and striving for static security, stability, a permanent state of harmony. This longing is often expressed through a static notion of "happiness" as a goal, or a state, which can be attained, a reference point or level which—in organismic analogies—is supposed to be sought and maintained through some sort of homeostasis, through adaptation to changes in the world around us. In the same stream image which I have introduced above, we would then see people, organizations, political systems, science, and whole cultures trying to win security by saving themselves on either bank of the river, gaining "solid ground" under their feet, shaping and perfecting the dry world around them, building and defending their "ecological niche"; or we would see them getting stuck on a rocky pinnacle in the stream, clinging to a particular view or model and interpreting from a fixed vantage point the movement of life continuing to flow past them. In both cases, security is gained at the expense of movement, of participation in life—the more recent history of science, favoring the "objective," the generalizable, the reproducible and the testable, even turned nonparticipation in the life stream into a virtue. In a later chapter, I shall discuss what can be done about this attitude of science.

Steering a course *in* the life stream requires awareness of its flow between opposites. Cutting across the stream, we may identify such opposites for any number of dimensions:

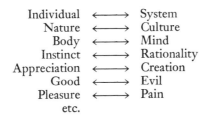

Human life is not an expression of either one or the other, or of a specific shade in between. In this book I am not endorsing the dualistic principle so deeply embedded in Christian thought—with life on earth to be rewarded in a neatly separated heaven. Rather, I see human life drawing its energy from *living out, as fully as possible, the tension* to which it is subjected in the field of force between such opposites. Any possibility for action depends on the existence of a potential, of a field of tension, between points in space and time.

Some of these opposites have been fiercely debated from one or the other point of view all along the history of human thought. Others have not been recognized as opposites at all and have been conceived rather as accumulative notions "on the same side," which removed the tension between them and eliminated them as a source of specifically human energy. This happened, for example, when love was placed atop sexuality and put in charge of control of the latter. *Eros*, the living out of the tension between both, and the greatest source of human energy for the antique Graeco-Roman culture, has thereby been driven out of the Western world or has become an empty notion confused with either sex drive or sublimated love. The Christian notion of *agape*—boundless, universal compassion—although related to *eros*, could not remain a viable concept in the absence of "the other side," sexuality, which became repressed in Christian thought. Other opposites, such as pleasure and pain, have been conceived as unambiguous alternatives of an "either/or" type and have even been used as a basis for some human or social cost/benefit calculations. The release or reduction of tension has even been made the principal factor in human behavior as misunderstood by a school of behavioral thought which dominated in the 1950s (especially N. Tinbergen). Whoever has listened to music or a poem with wide open heart and mind, or has watched a mime express the complexity of even the simplest, most basic of human feelings such as joy or sadness or love, knows that human life is not essentially concerned with seeking pleasure and shunning pain, but that it receives energy from living them both at the same time. Hölderlin, in his Sophocles distich, gives deeply felt expression to this unifying principle of humanness:

> Viele versuchten umsonst, das Freudigste freudig zu sagen,
> Hier spricht endlich es mir, hier in der Trauer sich aus.
> (Many seek vainly, joyously to express joy.
> Finally I apprehend it, here in my sorrow.)

The arts still convey to us a wholeness embracing these opposites, where formal language fails and science has become a narrow trap of onesidedness.

Steering a precarious course in a fast-flowing stream while drawing energy from the fields of force extending across it—this is now the image of the human predicament which I propose to use at this stage—obviously calls for regulation to cope with the uncertainties and complexities of the voyage. In other words, we need a policy to stay in the stream without drowning. However, again and again in human history the formulation and enforcement of policies took

the involved individual or human system *out* of the stream—and
stranded it on either bank, or pinned it up on a sharp rock. The
succession of longer- or shorter-lived cultures in history may be
seen as the alternation between getting stranded and struggling to get
back into the stream again—with quite impressive periods of steering
successfully with the help of policies which were "right" for the
conditions in a certain phase of history, but became "wrong" when
the conditions changed. We are at such a point in mankind's evolution where changed conditions invalidate all our policies that have
been so successful even in the recent past, and that presumably have
constituted the ideal response to a presumably unchanging and un-
changeable human condition. No wonder we are stupefied and confused—but our mistake is the same which many cultures have made
before us, namely to force a rigid model upon a fluid reality.

This recognition is not so difficult to accept if we take a quick
look at some of the policies that have been, or are, in effect for
human systems of one of the highest orders, that is, for cultures. All
of them have been successful in certain aspects and for some periods
—yet they differ widely in their basic assumptions. It seems to me
that there are at least four basic assumptions underlying every explicit or implicit policy, which in turn give rise to a host of further,
partly overlapping assumptions.

The first assumption concerns the basic or most "natural" state of
a human system. Is freedom—our general notion of a "good" order
in the system, permitting human capabilities to be lived out and become effective—to be assumed as given, as something to be guarded,
preserved and defended? Or is it something to be continuously built
and rebuilt by man as a human or social artifact? In other words,
is history a process of deviation from a perfect static cosmos and an
originally given "good" state of human systems, a process which
poses the challenge of *corrective action*—of negative feedback action,
to use the technical term of control theory—which strives to bring
the system back to an ideal state presumably implicit in its design?
Or is history essentially a process of self-organization of human
systems in an unfolding dynamic cosmos, posing the challenge of
creative action—of positive feedback, or feedforward—continuously
unsettling system states in order to create more ideal states? In the
following chapters, the same question will reappear in considerably
greater complexity. We may, for example, ask whether an assumed
"good" state is defined merely by the design of internal relations
within a system, or whether it also (or even primarily) refers to

the external relations between a system and its environment—in the latter case, corrective action would be guided by *homeostasis*, by adaptation to a changing environment so as to ensure the viability of relations between system and environment (the survival of the system in question). Thus, the basic assumption of a given order does not necessarily imply that this order is inherent in the design of the structure of a system; rather, it implies that it is inherent in the design of the life of a system, in other words in the design of its policy. In an evolutionary perspective, the same question will pose itself at an even more profound level.

The second assumption concerns man's role in designing and carrying out policies for human systems. Here, the most essential notion is whether this role is assumed to be an *amoral* or a *moral* one. This is the question that touches the innermost core of the human condition. Can we bear our freedom, which means can we bear our responsibility for it? Dostoyevsky, in *The Brothers Karamazov*, has given a pessimistic answer. The Grand Inquisitor addresses Christ, who has returned to earth and preaches freedom, with the following words:

I tell Thee that man is tormented by no greater anxiety than to find someone quickly to whom he can hand over that gift of freedom with which the ill-fated creature is born. But only one who can appease their conscience can take over their freedom. In bread there was offered Thee an invincible banner; give bread, and man will worship Thee, for nothing is more certain than bread. . . . But what happened? Instead of taking men's freedom from them, Thou didst make it greater than ever! Didst Thou forget that man prefers peace, and even death, to freedom of choice in the knowledge of good and evil? . . . But didst Thou not know that he would at last reject Thy image and Thy truth, if he is weighed down with the fearful burden of free choice? They will cry aloud at last that the truth is not in Thee, for they could not have been left in greater confusion and suffering than Thou hast caused, laying upon them so many cares and unanswerable problems.

The history of science over the past three centuries, and in particular the more recent history of social science, may be viewed as a continuous attempt to evade the moral issue, to tend toward an image of a mechanical universe populated by clockwork societies and human beings "beyond freedom and dignity"—in short, toward the "intellectual barbarism" denounced by P.L. Berger (1963) and "demonstrated admirably in the recent history of behavioristic psychology." Behavior is comfortably amoral, while action bears the moral burden. But the moral burden again becomes alleviated if action is just seen as a "trigger" to set off movement along a predetermined course, as a prominent version of Marxist theory holds, for example. Later in this book, the amoral stance of current science will be discussed in more depth, and possibilities for moral modes of inquiry will be explored. Moral choice is more than just the preparedness for deliberate action; it implies also accepting the responsibility for the consequences of that action. Moral action is inherently nondeterministic.

The third assumption, too, concerns man's role in that complex business of designing and executing policies for human systems. Is this a *collective* or an *individual* task? The whole big issue of "participation," so prominent in our time, lingers behind this question. But it becomes of even more fundamental importance when it is seen in the context of that particular version of the first assumption which stipulates that freedom is to be designed and built by man. We know that individual creativity works and, to some extent, even how it works. But is there anything like a social, a collective, creativity? Some animal species exhibit astounding examples of collective creation—ants and bees, to name only two with their outstanding architectural archievements; but it is a preprogrammed creativity, the execution of patterns of activity implicit in almost rigid genetic templates. Does man's inventive creativity in the individual sphere find an analogue in the social sphere—can it find one, in principle, or is this a contradiction in terms? I shall return to this question later. For the moment, the basic question is introduced here primarily because it gives rise to vastly different kinds of policy concepts.

Finally, a fourth assumption enters in a prominent way. It concerns the primary focus of the policy, what aspect is to be "optimized," or at least emphasized first—a *holistic* (systemic) or an *atomistic* (individual) aspect; society's needs and aspirations or the individual's? To formulate this issue in terms of monism versus pluralism would perhaps be slightly incorrect since a systemic focus does not necessarily imply a monolithic attitude toward life in the system. Since this fourth assumption is geared to the *purpose* of a policy, it is not surprising that it brings in vast differences in values and ideas about the ultimate purpose of human life. The chasm between socialism and capitalism, so conspicuously in the foreground of interest today, expresses itself primarily in the difference in this fourth assumption—with the other three assumptions identical in both cases. We may also say that a holistic focus is chiefly interested in the survival and well-being of a human system, or mankind in general, whereas an atomistic focus is interested in the survival and well-being of the individual. Thus, the difference between a long-range dynamic, or even evolutionary, perspective, and a short-range or static perspective is partially implicit in the span provided by this assumption.

One might extend this list by many more basic assumptions. For example, is the policy supposed to be valid in an absolute way or should it be flexible, contingent to situations? Does it assume a linear course—with a beginning and an end and a life of "progress" in between, as Christian religion assumes—or a cyclical development, or other cosmological patterns? Some of these assumptions will be discussed in the following chapters. For the purposes of this chapter, however, the four assumptions outlined above provide a sufficient variety of policy "models" to draw some preliminary conclusions.

As in almost all tasks of structuring complex concepts or situations or problems, it is helpful to employ *morphological analysis*. This is a widely used technique—in technological and also to some extent in social forecasting in a static framework—which tries to identify the basic parameters of a complexity and then proceeds to list varieties of qualitatively different versions which each parameter may acquire (Zwicky, 1969). The four basic assumptions have been discussed above in terms of two qualitatively different versions each. This is not meant to preclude the possibility of any "mixed," or watered down shades between the two poles—in fact, most of the assumptions underlying current policies seem to represent such slightly "impure" shades, nevertheless leaning more toward one or the other side. The basic assumptions may now be arranged in a four by two morphological matrix which yields 16 possible combinations (see Table 1). Each of these combinations may be readily identified with policy ideas with which any Westerner is today familiar. Many more may be identified with the basic structures laid out in this matrix.

The left-hand column implies principles which are aspects of a more passive life—of the yin—and the right-hand column principles which pertain to a rational and active life—the yang—thus emphasizing aspects which are specific for human life. However, both sides have to be seen again as opposites between which flows the full evolutionary stream, the tao. Balance may be found in the tension

TABLE 1

*A morphological representation of policy concepts,
structured by four basic assumptions.*

MORPHOLOGICAL MATRIX

A. Freedom is:	1. Given	2. To be built
B. Man's role is:	1. Amoral	2. Moral
C. Design/support task is:	1. Collective	2. Individual
D. Policy focus is:	1. Holistic	2. Atomistic

Examples:

A1—B1—C1—D1	:	Self-regulation by social instinct
—D2	:	Anarchist ideal (Communist ideal)
—C2—D1	:	Socialism in our time (totalitarianism)
—D2	:	Rousseauean ideal, capitalism, "free enterprise"
—B2—C1—D1	:	"Urbane" government (medieval city)
—D2	:	Welfare state
—C2—D1	:	"Enlightened" dictatorship (absolutism)
—D2	:	Paternalism, feudalism, mythical legitimacy
A2—B1—C1—D1	:	Maoist ideal
—D2	:	Marxist (revolutionary) ideal
—C2—D1	:	Technocracy
—D2	:	Meritocracy
—B2—C1—D1	:	Socialist ideal (modern communes, kibbutzim)
—D2	:	Democratic ideal
—C2—D1	:	Leadership, elitism
—D2	:	Original Christian ideal

between both sides. But most policies try to avoid this tension, to resolve the assumptions to one or the other side and even in a pretty wild mixture, as Table 1 exemplifies.

Such a morphological representation reveals some of the most amazing affinities between basic policy ideas, which are supposed to constitute extreme opposites but are opposites only in one or two particular assumptions, while sharing the others. Take, for example, the pairs capitalism/socialism, anarchism/free enterprise, christianity/feudalism, technocracy/socialism. Since each of the four basic assumptions is supposed to be essential, a difference in each of them may, of course, be considered essential—but only in one of several essential dimensions. It is the specific point of view taken which determines whether such policy concepts are considered as basically related or inimical to each other. Who would have expected such closeness between some of these conflicting notions—and how many people would be prepared to discuss some of the world-shaking conceptual clashes of our time in such structural terms? Human history provides many examples of such clashes, which have often led to bloody confrontations and major wars; one needs only remember the religious wars which sometimes expressed conceptual differences even within the same major religion, such as Christianity. To us, the differences underlying the Thirty-Years' War may seem relatively trivial—and to many of us also the current religious conflicts in Northern Ireland. By the same token, most of the conceptual differences which divide the world today may look trivial to generations to come—more so, since they may be viewed against the historical background of a mounting world crisis of unprecedented dimensions.

Other affinities seem to suit our traditional view in a better way; for example, democracy/welfare state, technocracy/meritocracy/capitalism, elitism/dictatorship, modern communes/medieval cities. We have few qualms about these relationships among concepts. However, some of these affinities make it clear how utterly precarious is the life of a human system, as guided by a particular policy, in a multidimensional field of tension. If leadership is so close to dictatorship, what can we do to avoid any "flipping over"—and can the boundaries between the two be clearly drawn at all? Does democracy always end up in the welfare state, as some European examples —the Scandinavian countries, Great Britain, Austria among others —seem to suggest and the contemporary trend, in the United States as well, seems to confirm?

The pairs that differ only in the first assumption; that is, the opposite numbers of the A_1 and the A_2 groups, deserve special attention. The A_1 attitude is basically a static one, resistant to change, focusing on structure, prone to satisfy the basic human need for security and stability. The A_2 attitude, in contrast, is basically dynamic, seeking change—whether through "blind" movement or design (B_1 or B_2)—focusing on historical processes rather than structure, and expressing the human need to strive toward "a better world." Where human energy becomes organized toward a specific direction in the A_2 case, creating vectors of movements or, in other words, history, energy is largely unordered in the A_1 case. In terms of a thermodynamic analogy, we may say entropy is high in A_1 and low in A2—but these considerations will be elaborated more clearly in Chapter 4 when the characteristics of human systems will be discussed.

Suffice it here to state that assumption A1, in most cases, tends indeed to downgrade the life of a human system either to the lifelessness of a closed mechanical system or to the passive life of a biological system within natural ecology—in both cases, the ultimate consequence is rigid control by an imposed or genetically fixed template, to which cultural templates become subordinate. This rigidity often makes itself felt, not so much through a compulsion to act in a certain way, but through frustration in not being able to act in a chosen way. It is this frustration which gives rise to a growing alienation from "the system" which, in many cases, through its institutional framework, is doing nothing more than just trying to continue a traditional policy (given perhaps in the form of a religion, or a constitution, or an ideology). Naive and romantic ideals of freedom, as they express themselves, for example, in such concepts as "free enterprise" or "anarchism," ultimately become reversed if the A1 state acts as a trap from which the human system cannot find its way out.

Human systems, like human beings, tend to jump to an A2 state in times of energy reception—often in major quantum jumps, such as cultural mutations or revolutions—and to slide gradually toward an A1 state in times when they use up their free energy. The meaning of human systems "receiving" and "spending" energy will be clarified to some extent in the following chapters. What I want to point out here, however, is a certain basic tendency inherent in all A2 configurations to move toward their corresponding A1 configuration—which does not preclude that sometimes the development ends up somewhere else, as is evident from the fate of the Communist revolution, which is no longer seen as a movement toward ideal communism but is caught in the rigid totalitarian and bureaucratic patterns to be found today in countries which identify themselves as socialist.

It is particularly interesting to recognize, in the patterns set by Table 1, those policy configurations which are operative in Western-type societies today—not only because it concerns our own culture, but also because we find a curious mixture of concepts, some of them even in apparent conflict with each other, such as technocracy and "free enterprise," democracy and Christian belief, or capitalism and welfare state. Such a conflicting array of policy concepts, instead of a shared central concept, may be considered as a symptom of the pathology of cultural systems; a symptom characteristic for an "end-game"—from which will spring a new approach, a new culture.

But our Western culture is also interesting because much of it has been shaped and originally determined by the Christian religion. In Table 1, the "pure" idea of Christianity occupies a position at the extreme opposite of a self-regulating ecology of a natural type. In other words, Christian belief incorporates the utmost challenge to develop man's rational capabilities. Not all religions were so imprudent as to show the way to their own ultimate destruction. Some of them, like Buddhism, have avoided making any social assumption under A at all. Christian religion, one might say, took an active A_2 position and placed transcendental promises mostly into the D category of assumptions—to state, with Table 1, that Christian religion implies an "atomistic" D_2 assumption would therefore not be fully correct, since it is not the individual's life on earth which is to be optimized but "life after death." What matters most for policy, however, is the Christian challenge to win life after death not just through contemplation or development of the self (as in other religions) but through "progress"—even linear progress toward an ultimate Doomsday—and through achievement in building a social system on earth to reflect the order awaiting man in heaven. More than in any other recent religion, the ideas of balance and regulation are absent from Christian belief; on the contrary, the transcendental promise is built into it in such a way as to stimulate and excite action on earth through a central myth of insatiable expectations which suppress any concern for balance in and alongside this action, that is *on earth*. Christianity is not to be blamed entirely for the degeneration of this myth, and for all the material growth syndromes and damages to our environment which, to some extent, have grown out of it. But a religion which is in itself dualistic and unbalanced between heaven and earth is bound to lead to serious trouble as far as earthly matters are concerned.

Where, then, does balance reside in policies which are supposed to guide the lives of humans and human systems, lives which are led on earth but in a transcendental perspective? As the central theme of this book, this will be discussed under various aspects. A few notions may already be formulated at this point, and I shall advance them in the next chapter—notions which represent a highly subjective selection. After all, the evolving arguments in a book have a life of their own; and this life is necessarily linked with the life of the author. I do not make any claim of "objectively" revealing any "truth." The brief empirical excursion in this chapter may be deceptive. From here on I shall develop my subjective view of how balance may be pursued and policies designed for healthy lives of human systems.

(2)

ON BALANCE AND CHANGE

In the previous chapter I pointed out that almost every kind of policy—every kind of model for regulating human systems—has been proposed, and many of these models have actually been, or are being, applied. Some of them even were applied side by side within the same cultural setting and what appears to be essentially the same historical situation. Does this mean that some policies are "right" and others are "wrong"? And what does "right" and "wrong" mean in this context?

More specifically, we may first ask whether there is one "right" policy for mankind, valid at all times and for all parts of humanity, or whether there is a variety of situation-contingent "right" policies. Messianic religions and ideologies usually assert the one-right-policy view since, for them, the purpose of policy is to be seen in directing human systems toward a prescribed endstate—usually a tensionless state on earth, such as "the classless society," or the equivalent in heaven. Therefore arises the zeal with which messianic myths defend their corresponding policy concepts.

Obviously, one enduring policy won't do. Human history has seen a never-ending succession of cultures—each incorporating a specific basic policy concept—come to life, grow and bloom, dry out and wither away. We are even so aware of the temporary character of cultures that we compare them to biological organisms which, necessarily, have to die—an interesting, but basically deceptive image. There was an obvious need for different policy orientations in the different phases and dynamic situations of mankind's psycho-social evolution. Since the phase in which we are living now is the most dynamic one mankind has ever known—at least as far as the most influential part of mankind is concerned—and changes involve ever shorter time factors, we must become particularly aware of the demands made on us in this evolving situation.

As if this were easy. Rational planning strives to introduce such a flexibility in concrete options in a framework of thought and at a level which is called "strategic," a notion to which I shall return in later chapters. Organizational principles facilitating and enhancing the potential for conceiving as well as implementing a basic flexibility in the choice of operational targets have been formulated and tried out. But nobody, at least in modern Western society, has yet tried to be equally flexible with respect to policy. On the contrary, our whole institutional fabric functions with the explicit or implicit aim of keeping firmly to a given rigid policy. Governments everywhere are supposed to preserve constitutions; churches their respective religions; and political parties their ideologies; just as business is meant to serve economic growth, and academia is expected to defend a particular model of science.

Western society has been remarkably successful in the application of policies which have greatly enhanced the lives of individuals as well as the lives of political and economic systems, of institutions and organizations. But it is dawning on us rapidly, now that regulation in accordance with the old policies is breaking down, that the activities of humans and human subsystems are clogging and blocking the life of the larger human systems, and indeed, of the entire world. These activities obviously lack balance in the greater context. It is not my intention to review them here and to present one more survey of the flood of predictions, contingency forecasts, and simulation studies which are but expressions of a widely shared and deeply held "gut-feeling" that something essential in our human world has gone "off center." I am deliberately using these unscientific notions, because I shall return to them in this book and deepen their meaning. For the moment, suffice it to point out, with Vickers (1970), that regulation is breaking down in at least four areas: ecology—or the relationships between ourselves and the physical environment; economy—or the relationships between ourselves as producers and as consumers; politics in the broadest sense—or the relationships between ourselves as doers and as "done-by"; and appreciation—or the consistency of our values, criteria, and measures by which we judge performance in human tasks, including the task of regulation.

In the Western world, the breakdown of regulation may be blamed, at least on the surface of things, on a focus in policy which is too individualistic. Equally, in the so-called socialist countries the grinding difficulties with regulation may be blamed on the one-sided emphasis on systemic focus. "Very well," somebody may say, "we can always change one of the basic assumptions underlying our policies; for example shift from D_2 to D_1 in the matrix shown in Table 1, from an atomistic to a holistic focus, and change over from a democratic to a socialist ideal, from the welfare state to an 'urbane' form of government." This would, in itself, not seem so bad. And modern socialism, by shifting from D_1 to D_2, would then change into capitalism and free enterprise. But this is absurd—juggling with opposites would just reverse some of the ideological frontiers and not bring much change for the world as a whole. Besides, these are incisive changes which affect much more than just one aspect of policy, and therefore they will not happen at the snap of a finger.

What is already happening instead, under the pressure of the evolving situation, is a movement toward the center between opposites. The free-enterprise world is moving in the direction of increasing recognition of a shared responsibility for the larger human systems, for society-at-large; and modern socialism discovers the power of a reward system geared to the individual and human entrepreneurship. Both sides are actually in the process of moving toward some balance between the opposites of assumption D concerning the primary policy focus.

However, flexibility in policy may come more naturally to a non-Western mind. Was Mao Tse-tung's "cultural revolution" not meant to soften, or even break up, the encrustations in the management of the country and the party all the way up to the level of policy?

The *I Ching* (the old Chinese Book of Changes) gives a pattern of 64 hexagrams, each of which stands for a particular basic policy, applicable to the life of an individual or a human system. But more importantly, it structures the transitions from one to the other; it presents a network of change processes rather than a structure of policy models. What this means was brought to me in a most impressive way, and with frightening clarity, when I threw the *I Ching* for the first time and asked for guidance on my own personal policy—although I then called it "dharma," which seemed somewhat more congenial in the context of age-old wisdom.

I did not ask the *I Ching* in a playful mood; it was after an intensive Sunday morning meditation with my friends of the Kundalini Yoga ashram in Berkeley, when I felt the moment to be "right." The hexagram I threw was called "After Completion," the only one in which all lines are perfectly in place. "The transition from confusion to order is completed," stated the old commentary, "and everything is in its proper place even in particulars. . . . This is a favorable outlook, yet it gives reason for thought. For it is just when perfect equilibrium has been reached that any moment may cause order to revert to disorder." The image of "After Completion" is water over fire, two elements hostile to each other, but generating energy when brought into balanced relation. "If the water boils over, the fire is extinguished and its energy is lost. If the heat is too great, the water evaporates into the air. . . . Only the most extreme caution can prevent damage. In life, too, there are junctions when all forces are in balance and work in harmony, so that everything seems to be in the best of order. In such times only the sage recognizes the moments that bode danger and knows how to banish it by means of timely precautions." The way I threw the hexagram "After Completion," it changed over to a hexagram called "Abundance" (Fullness)—referring again to a development that has reached a peak and thus suggesting that this extraordinary condition of abundance will be brief. But: "Be not sad. Be like the sun at midday" was the advice given by the old commentaries. It seemed like the ultimate confirmation of my life of change which had become increasingly fluid in its themes and values—its policies.

In our search for viable policies—and having discarded the one-right-policy attitude—we may now ask what constitutes a better approach to living in the presence and under the influence of opposites; trying to attain some flexibility in the alternation between policies made up of various combinations of onesided assumptions—such as the examples given in Table 1—or trying to design policies based on assumptions which deliberately bring into play the tension between opposites and without attempting to resolve it onesidedly. Returning to the image of the life stream, which I introduced in the first chapter, we may view this alternative in a graphic way: Should we try to cross the river in both directions over and over again, resting in between in the security of one or the other solid river bank—but never for too long, and never settling down for good; or should we rather try to continuously stay in the stream and never leave its invigorating, fast-carrying flow?

We are still facing what appears to us to be a basic dilemma between opting for a secure and perfectible but static world—a monument to life rather than an expression of it—and a risky and elusive but dynamic world in which our fate evolves with the forces of life in and around us. The dilemma is within ourselves, the clash between images of the world which we conceive. It is the oldest dilemma in the history of subjective human thought. Its beginnings are veiled in the fog enshrouding the origins of cultures.

But it seems that evolutionary thinking, as it later expressed itself in various cultures, goes back to the teachings of Hermes Trismegistos, as they have come to us in the *Corpus Hermeticum*. They convey an evolutionary picture in which a static All generates a succession of universes which follow evolutionary courses, are born, evolve, and die. Whether Hermes was a historical figure or not, remains a controversial issue. Also, we do not know in which period he lived in Egypt—presumably long before the days of Moses. The Egyptian god Toth, as well as the Greek god Hermes, are said to perpetuate Hermes Trismegistos. The earliest preserved written texts of Hermetic philosophy, in Greek language, seem to stem from around 100 A.D. But Hermetic philosophy seems to have acted over the millennia as a kind of transcultural archetype, as the fertile ground from which sprang the evolutionary world views in Mediterranean as well as Asian cultures, Medieval alchemy as well as some of the more recent nondualistic mystical and philosophical schools.[1] It postulates the evolutionary principles which now are gradually coming into the focus of scientific attention, as I shall outline in the following chapter.

1. In spite of the development of science from a predominantly static perspective, Hermetic philosophy was held in high honor until science became one-sidedly materialistic three hundred years ago. It is interesting to note, for example, that Newton, who laid the theoretical foundations for a modern static cosmos, tried to win recognition from his peers by pleading with them that his theory was in the spirit of Hermetic philosophy! In the nineteenth century, Herbert Spencer and some writers in the English and French occult tradition revived the principles of Hermetic philosophy. A compilation of Hermetic axioms and corresponding comments by three anonymous "initiates," published in 1904 under the title *The Kybalion*, anticipated the evolutionary perspective which physical science started to acquire one year later with the publication of the theory of relativity. In our century, the philosophy of A. N. Whitehead is perhaps most representative of a modern Hermetic view. What for Hermes was "the All," is now Whitehead's "extensive continuum"; both notions pertain equally to the generation of evolutionary dynamics in the universe.

A recently proposed cosmological paradigm (Tryon, 1973) views evolution as a "vacuum fluctuation" of the void, as it is to be expected from quantum field theory. Such a view is capable of explaining the beginning and the end of an evolutionary arc of energy processes which springs from the void and returns to it without leaving a trace, since a vacuum fluctuation is symmetrical in all physical characteristics so that the sum is always zero. This seems to point to the concept of a simultaneous evolution of two nonequilibrium universes, one manifesting itself through matter, the other through antimatter, which will eventually annihilate each other in an equilibrium void, or "All."

Perhaps man's basic, still instinct-guided, world view was evolutionary. For Hesiod, at the beginning of Greek philosophy around 800 B.C., the universe was still dynamic. The great schism came with Parmenides, in the fifth century B.C., who conceived a static, mathematical cosmos with which we are still dealing in contemporary science as well as in many other aspects of today's world, a cosmos created to perfection and balanced in this perfection. The word "cosmos" originally meant order, and it was interpreted as a static order. Such a static cosmos would swing in cyclical movements but always remain itself in all eternity. All order is already built into it and can be deduced by unraveling the order of the universe. The static concept thus extends also into man whose inner order is a reflection of the cosmic order. Although Greek tragedy gave magnificent and profound expression to an evolutionary concept, it did not lead to a general reflection on evolution in Greek philosophy. But Heraclitus conceived the *logos* as the evolutionary principle and stipulated his famous "panta rhei" (everything is in flux); time was assigned a direction, but this did not yet make the static cosmos crumble. Plato took a compromise position; for him, the cycles had a beginning, a structure and a meaning. The Judeo-Christian religion assumed a world created in six days with ready-made species, including man, but not to perfection—history here is the unfolding of a divine plan, a plan with a time dimension stretching toward Doomsday, a possibility and an obligation for improvement on earth and in the universe. The Newtonian world still is an expression of the Greek cosmos of mathematical structures. The evolutionary approach which destroyed the Newtonian static world, really started with Kant's insight that "creation is not the work of a moment" and that "creation is never complete." His booklet *The Nature and Theory of Heaven*, published in 1755, from which these quotations are taken, became the starting point for a modern concept of time.

In an evolutionary view, the *event* is reality, not any structure such as space or matter. In the event lies the unity of space, time, and matter. Every movement has its own time built into itself; instead of metaphysical time we have physical, biological, or cosmological time unfolding in the corresponding processes of evolution. This grandiose insight became more widely accepted in biology than in physics. Darwin published "The Origin of Species" in 1870; Bergson introduced his concept of an "élan vital," a life force, working in an antientropic direction toward higher forms of organization, in 1905—the same year in which Einstein's theory of relativity shattered the static Newtonian world and introduced to physics the evolutionary focus on the unfolding of a space-time continuum. As A. Katchalsky[1] remarks, the theory of relativity should really be called the theory of the absolute—in it, the event is the absolute, independent of the setting. Up to our time, this evolutionary concept of the physical world has not yet penetrated very deeply into science. It is only very recently that attention has been drawn to the possibility that the laws of nature themselves— the mathematical cosmos of the Greeks—may undergo evolutionary changes.

In an evolutionary world, there is nothing to be learned merely through deduction and induction, but all through *operation*, through an evolving movement and the experience streaming from it. It is a dramatic world in which roles are acted out—roles of atoms and molecules, of organisms, of human individuals, institutions, and cultures—and in which a great play unfolds itself, greater than mankind, greater than life itself.

In a static world, the relations between an individual and society would be recognized as being rigidly and hierarchically determined. The organizational pyramid which we still find in many organizations as well as in society at large, corresponds to this view. It signifies order of a *structural* type, with processes (often of a steady-state, or at least a qualitatively unchanging type) subordinate to it. In a dynamic world, the relation between an individual and society would be viewed as being basically part of a horizontal network as well as of a hierarchy—human life radiating in all directions and animating the life of human systems through which it would find enhanced expression of an innate kind, not of a remote "higher order." What counts here in organization are the *processes* first, with structure kept as fluid as possible.

In a time of rapid qualitative change, such as ours, a dramatic world becomes virtually forced on us, to allow for a freer play of all the innovating forces bursting forth and which ultimately will explode fixed structures attempting to contain them. But this means

---

1. I am indebted to the late Prof. Aharon Katchalsky for some details of this historical overview which he pointed out in the Berkeley seminar on "Time," held in 1970.

that first we have to test the human capacity to animate and carry such a dramatic world.

To what extent can we become flexible—and ultimately fluid—in the regulation of our personal lives and the lives of our social and cultural relationships and systems? What degree of "surrender" to the life stream, of uncertainty and complexity are we apt and willing to bear? These are not easy questions to answer. Our responses to them depend to a large extent on our image of what constitutes human nature, and on our active acceptance of that image.

In the second part of this book I try to sketch those aspects of human nature which bear on our playing an active role in evolution —aspects which I shall then explore in the rest of the book. At this point, however, I should like to briefly outline the general nature of the processes with which this book primarily deals. They may be characterized by the notions of temporal and spatial self-balancing. The term "space" is used here in the general topological sense and denotes not just physical space but also, for example, social and spiritual space.

Temporal self-balancing of a process may be viewed as a process which is always linking itself back to its own origin, and thus becomes "aware" of its own direction and position in space and time. It proceeds by prancing back and forth in feedback loops, and feedback loops within feedback loops within feedback loops.

Bergson had already pointed to the formation of memory through reiterated interpretation in such feedback learning processes. Even microorganisms, in probing their environment and selecting "good" directions for their next movement, seem to develop a memory and the capability of comparing. But such a process as the one referred to above, does more than just summarize a memory, a learning experience; it operates, it ventures out to realize its innate potential, and its form. It does not just seek peace with the world, or some kind of homeostatic relationship—it acts upon the world and stakes its own claim. Although there may be general guidelines for the overall direction, such as a variational principle to be "checked" with every step taken, the process seeks its own way and finds its justification within itself, not in any external regulator or guarantor. This introduces an important difference from a static view which needs external guarantors of systems, and ultimately an external god. Here lie the roots of the dualism which rules Western concepts of the world and man's position in it.

In an evolutionary view, the linking backward—the *re-ligio* in the original meaning—reestablishes the unity with the origin, with the wholeness of meaning, and thus continuously enriches the process which, by necessity, acts out only partial aspects of this meaning. The well-known biogenetic law of ontogeny recapitulating phylogeny—the growth of the embryo following the same pattern the development of the species had followed—is the expression of such a re-ligio. And when we speak of religion in the spiritual sense, we mean this same unity in the link backward, bringing out, as Whitehead has put it, the eternal in the temporal passage of facts. The evolving reality of such a self-balancing, self-nourishing process has been beautifully evoked more than twenty-two centuries ago by the Chinese poet Chuang-Tzu (demonstrating, at the same time, the evolutionary orientation of ancient Chinese philosophy):

In the Beginning of Beginnings was Void of Void, the Nameless.
And in the Nameless was the One, without body, without form.
This One—this Being in whom all find power to exist—
Is the Living.
From the Living, comes the Formless, the Undivided.
From the act of this Formless, comes the Existents, each according
To its inner principle. This is Form. Here body embraces and cherishes spirit.
The two work together as one, blending and manifesting their Characters. And this is Nature.

But he who obeys Nature returns through Form and Formless to the Living.
And in the Living
Joins the unbegun Beginning.
The joining is Sameness. The Sameness is Void. The Void is infinite.

In Chapter 5, I shall describe processes of artistic and scientific inquiry as such self-balancing processes realizing innate forms.

Spatial self-balancing of a process again works in feedback loops, not just in "big strokes," but again through an intricately interwoven pattern of feedback loops within feedback loops within feedback loops. There is no preferred spatial direction here, since all processes with which I deal in this book are two-way feedback processes. The processes, of course, unfold over time, so that in all spatial balancing there is also the dimension of time and, in an evolutionary view, also the direction of time.

The basic symmetry in the processes underlies pairs of opposite notions such as deductive/inductive, appreciative/creative, feminine/

masculine, or passive/active, which in the end all boil down to the
same basic notion of yin and yang, the taking and the giving in
Chinese philosophy. Each side contains the seed of the other; each
side becomes real only in terms of the process linking it to its opposite. All creation follows this pattern, whether in evolution at
large or in purposeful human action—in design. Science prefers to
speak of deduction, the process of arriving at particular conclusions
from a given order (a set of assumptions), and of induction, the
process of arriving at general conclusions (a new order) from
particular observations. In human affairs it is perhaps more suggestive, as Vickers (1969) words it, to speak of appreciative and
creative phases which both "change the norms to which they consciously or unconsciously appeal. The appreciative phase changes
them by the mere fact of using them to analyze and evaluate a
concrete situation; for this may affect both their cognitive and their
evaluative settings. The creative phase affects them by presenting
new hypothetical forms for appreciation." The same type of self-
realizing and self-balancing feedback processes underlies my model
of the human design process which I shall outline in Chapters 6
and 7.

Spatial self-balancing of a system of processes, finally, takes the
general form sketched in Figure 1. There is no fixed element, no

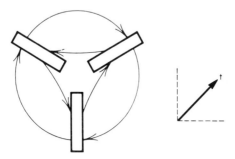

Figure 1. Spatial self-balancing of a system of processes.
Their interaction generates structures which may be
compared to standing-wave patterns rather than solid
constructs.

support from outside, no preference in direction or weight. If there
are just three elements linked by three pairs of feedback processes,
there is also no redundancy. This is the case for my model of the
human design process as well as for my general model of human

systems which I shall outline in Chapter 4. I do not know whether this expresses a profound underlying principle or is merely accidental.

Thus, systems of processes such as the overall design process, or human systems, balance themselves out in processes which are asymmetrically organized in time and—as I shall show for the design process—at least to some extent also in space. This is the meaning of evolution as I shall apply it in this book: as processes and systems of processes unfold with time, they evolve their own innate norms. They are also self-energizing. In an evolutionary perspective, the old image of change through external action—from a fixed Archimedean point from which it is possible to unhinge the world—is superseded by the image of Baron Münchhausen pulling himself out of a swamp by his own wig. Not a very encouraging image, true— but only as long as we insist that a given system (a man in a swamp) cannot by itself increase its order (separating the man from the swamp). In the following chapter I shall outline a few general ideas which have developed in topology and physics and which point to the reality of such self-realizing, self-enhancing processes generating mutations to higher-order regimes in given systems.

The issue is thus not just balance within a static cosmos but balance in progression, in movement. Balance in movement is not necessarily more difficult than balance at standstill—it may even be easier, as the example of the bicycle demonstrates—but it requires a different approach.

Dynamic balance may be achieved in two modes. People in urbanized Western society are used to a type of walking in which they continuously unbalance and rebalance themselves. The same happens with many forms of learning processes and the formation of personal relations which often imply giving up a secure position in order to gain a new one. But there is also a mode of walking in which balance is maintained at every moment. The body tries automatically to change to such a type of walking if it has to carry heavy loads. In rural societies, women are used to walking in this way while carrying loads on their heads. The resulting graceful movement may serve as a model for many activities in human life which gain gracefulness by maintaining balance in every moment.

When walking on the narrow rail of a railroad track—which all children love to do—balance can be maintained only by looking into a far distance; the moment one focuses on the next step, balance is lost. The same holds for the "gait of power" which Castaneda (1972) describes as he has learned it from a Yaqui Indian; this is a way of running fast through pitch-dark night with knees raised high and never a look to the side. I have had similar personal experience when, on several occasions, I was surprised by the fall of night before returning from a mountain tour. I virtually ran and jumped down very rough footpaths in black darkness three times as fast as it would have taken me to negotiate them in bright daylight. Under certain conditions, abandoning the reflective mode of action does not (as we might expect in our rational framework of thinking) lead to increase in randomness, but to superior order, to a new kind of balance with the surrounding forces. As I shall point out in the next chapter, nonequilibrium systems generally exhibit dynamic characteristics which are far from random.

Many people, especially the young in America, have become alert to two basic notions which guide many of them in their conscious personal development, regardless of the way they choose. There are psychological techniques such as encounter groups or gestalt therapy; physical ones such as structural integration and patterning; and spiritual ones such as Yoga and meditation. In Chapter 9, I shall come back to these to explore in more detail their significance for the ideas of balance and policy design. But I want to introduce just now the basic notions on which they all focus: *grounding* (or "becoming grounded") and *centering* (or "getting centered"). They form two parts of a triad, the third part of which has not been given an equally suggestive name and which stands for the mode of living which abounds today in Western society; maybe the slang expression "ego trip" catches best what distinguishes it from the other two notions.

The basic image of the life stream can serve again to help explain these notions. Western man has, so to speak, built himself a world on the dry "ego trip" bank, on the side of extreme rationality and artificiality. He shies away from the life stream, for he knows that entering it would be extremely hazardous for him since he has even lost sight of the opposite bank. This opposite bank is man's instinctive nature, his dark and evolutionary, even demonic, subconscious which goes back to his origins and is the source of his vitality. Western culture has suppressed this side which is buried in man's sexuality, and has driven man on ambitious and onesided "ego trips" in search of forms of spiritual satisfaction to which the great achievements of the human intellect—but also its atrocious aberrations and excesses —bear testimony. In a world increasingly reacting against the excesses of technology, the rediscovery of the opposite bank has become an urgent matter of survival. This process of rediscovering man's full and undivided nature is called grounding. It means to stand with both feet on the earth, to communicate with the energy streaming from it (from the mythological womb, from our Mother), to let oneself be grasped by the life force. A grounded man knows what he needs and what is good for him, because he knows in a very basic way, not dependent on logic, what is natural for him. And this —in an evolutionary perspective—is the all-important question which perplexes all rational creative planning and action today so that it trickles off into vague philosophizing. For there is no way to find out rationally and logically what is good for man and his systems —as much as we may strive for an empirical consensus. Static notions such as happiness have eluded us in spite of all our attempts to satisfy material needs and desires. The individual does not know what is good for him if he is not well grounded—and much less can the members of a human system determine rationally and logically what is good for the system.

Centering means, first, "putting one's center into oneself"—not anywhere outside, as into another person, an authority, society, or an impersonal mechanism of history. In the static Greek cosmos, man had to view himself in the physical center of the universe in order to feel centered. In an evolutionary perspective each process becomes centered in itself, as I have tried to explain above with the notions of temporal and spatial self-balancing and the self-realization of innate norms and forms. Centering would then simply be the steering of a viable course in the evolutionary stream in contact with both banks, the ego and the biological nature of man. But this is precisely what I also said of policy in Chapter 1. Centering, indeed, is the core of policy design—the design of a viable course for the life of an individual as well as a system. Its poetic image is the life of a flower which unfolds as a self-centering process between the roots, which are firmly grounded in the earth, and its aspirations which make it strive toward the sun.

Thus, we may retain the forces acting along the dimension

$$\text{Grounding} \longleftarrow \text{Centering} \longrightarrow \text{"Ego Trip"}$$

as one of the basic configurations characterizing human ways of relating to the world.

Psychology, and with it psychoanalysis, has developed a host of concepts which include the idea of grounding and is discovering that its scope cannot neglect centering. If Freud's approach may be characterized by a static adjustment to individual past history in the subconscious, and by a mental/libidinous focus, C.G. Jung widened it to encompass the enduring themes and cultural past history in the subconscious, with the focus on the spiritual side of man. W. Reich extended Freud's approach of static adjustment to the past to encompass now the whole body, with a genital emphasis, and started the rediscovery of a body/mind unity which the West had practically forgotten; the correspondence between features and distortions in man's mind and flesh opened up the possibility of developing physical techniques, grouped under the name of "bioenergetics," to enhance the awareness of this unity. A. Lowen went a step further and focused, not on adjustment to the past, but on achieving a body/mind balance in the present, in the much-publicized "here and now" mood—still a static balance, though. O. Rank and others saw their task in reinforcing the life force in all its manifestations and directions of flow, and thus brought in a dynamic perspective. With Keleman (1971), the notion of dynamic balance enters clearly. This balance is not just internal between body and mind, but also within man's natural and social habitat. With this opening up to the future, man's power to do things, his "ego trip," is allowed again, but in the perspective of a balanced attitude. Keleman focuses on the notion of gracefulness for the movement of a healthy, well-centered person—a notion which I shall adopt in this book because it encompasses so well the essence of policy in guiding the lives of individuals and systems, no less than the ways in which a living body moves. We are reminded here of Menander's verses in which he speaks of the enchanting gracefulness (*charis*) distinguishing man when he is truly human.

These are just new approaches to old themes, but they are timely, for the Western way of living has become grossly unbalanced, leaning heavily toward the "ego trip" side. Therefore, grounding is a prerequisite for gaining a feeling for the tension of life, from which springs a feeling for the meaning of centering. However, old cultures have at times developed a quite extraordinary sense for balance which they recognized as the core of individual, social, and cultural life. In this context I should like to point to the Japanese idea of *hara* which, literally translated, means *belly*, the center of gravity of the body (K. Dürckheim, 1956). Centering through hara actually applies to three dimensions; one extends "vertically" between the opposites of earth and heaven, of man's space/time reality and the divine beyond space and time; the second "horizontally" between the opposites of spirit and nature, of consciousness-increasing mind and the subconscious; and the third to movement in time. It is in the tension between the opposites, between the ego-fixing of the identity and the sinking down to the roots of life, that the unity of life is experienced, as it is expressed in the unity of body, mind, and soul. Dürckheim, a Western student of hara, states that "to regain awareness of this unity is precisely man's *raison d'être*, the nerve centre of human endeavour and the background of all his longings for happiness, fulfillment and peace." (Dürckheim, 1956).

I shall use the concept of centering in all three dimensions outlined above and extend it in Chapter 10. In this form, it underlies any possibility of tuning action in to evolution, the great stream in which life itself is embedded. In the next chapter I shall explore further the relationship between human design and evolution, starting with an outline of pertinent physical and biological concepts which are bringing the idea of evolution more and more into the focus of scientific and philosophical inquiry.

(3)

EVOLUTION AND REGULATION

Action is bound to a belief in the future; the mere enjoyment of the "here and now" can at best satisfy a behavioral perspective. Pleasure and the release of tension, whether promised by anarchist factions or by humanistic psychology, do not lead to the liberation which meaningful action holds. In a time in which man is conscious of his power on earth by way of interfering with natural systems and touching forces of life which he now recognizes as vital to himself, the need for a sense of purpose and direction is felt more urgently. This need arises in a phase of mutation in which the religious faiths underlying most of the existing cultures do not provide consistency anymore. History is no reliable guide across such a mutation—it pertains to the acting out of the beliefs held by a specific culture. Man needs a new conviction for himself, a new faith in his own future and in the validity of what he is doing with so much power.

The very recent surge of interest in long-range planning may be interpreted as resulting from this need for a sense of purpose and direction. But purpose and direction cannot be read out of feasibilities and options, projections and anticipated consequences, systems effectiveness and cost/benefit. Rather, they have to be assumed in order to make the whole instrumentarium, which rational knowledge has built itself, work at all. Work for which purpose and in what direction? These questions cannot be answered by patching together pieces of knowledge which, in the last end, have all been gained by deduction from a set of assumptions which we ourselves have made.

The usual response to this situation is to stipulate, if only subconsciously, that the basic constellations of the present will hold forever, at least the spiritual constellations of our cultural beliefs and values. Predictions of value dynamics still constitute the most unimaginative part of all forecasts; everything valued today by the "official" culture, is predicted to be valued even more highly in the future—more efficiency, more growth, more consumption, more competition, and so forth. Reality overtakes these poor forecasts even while they are being made.

On the other hand, so-called apocalyptic forecasts may frighten people by making them insecure, but they do not lead to any action to stave off the forecasted disasters. Continued growth is bad, granted—but what is to be done when Western culture, every aspect of it as it is engrained in individual and social life, is a culture with growth as one of its core values? Western statesmen listen to warnings concerning global limitations and the next day sign bills to ensure continued above-average growth. Japanese industry tells its government that it is all for limiting its growth—to ten percent per year.

The static order of the Greek cosmos is still with us and governs our theories of physical and social order in the world. That the spiritual order is crumbling can no longer be disguised, but the scientific disciplines, which are traditionally the custodians of physical and social order, disclaim any connection between their business and the scandal next door.

It took some twenty-five centuries for the idea of evolution to gradually erode the crystalline splendor of the mathematical cosmos. What is left of it, preserved in universities and the strongholds of economic determinism everywhere, will clash more and more severely with the dynamics of reality. The great idea that every process generates its own norms and measures as it unfolds—that the process creates the structure, not vice versa—has sunk its roots into part of modern science since Darwin, more than one century ago. But it is only in our days that a veritable confluence of insights, arrived at through scientific and mythical approaches alike, points to an impending synthesis of a general evolutionary view of the world. This flow of insights is stimulating the physical and the life sciences to develop new paradigms of far-reaching consequences—only the social sciences so far remain blocked off from it. As they face the challenge of dealing with an increasingly dynamic and changing social reality, an evolutionary perspective constitutes at least a step in the right direction, shaking off the old static and steady-state straitjackets. But I hope to present convincing arguments in this book that an evolutionary perspective of human and social life is more than an interesting hypothesis—is, indeed, part of the emerging new synthesis. An evolutionary perspective also ends the dualism between man and nature, the artificial and the natural.

Man with all his powers of design and realization, all his technology, is a manifestation of evolution as is all the rest of nature.

But evolution is more than change, more than an array of interacting processes. Where "process is an order of change," as Laszlo (1973 a) points out, "evolution is the coming into being of a new and higher order of process." It is this self-organization of processes, this ordering of orders of change, which is coming more sharply into focus as evolutionary theory merges with a dynamic general system theory.

The evolution of life is now firmly established. So seems to be the evolution of matter, from the transformation of atoms and molecules all the way to stellar evolution and the evolution of whole galaxies and the entire universe. But the basic assumptions governing the organization of evolutionary views stem from the static cosmos of equilibrium or near-equilibrium systems. The most important consequence of assuming equilibrium conditions is the Second Law of Thermodynamics which states that the world is irreversibly and *monotonously* approaching a state of universal disorder, or maximum entropy as one says in physics. In a closed system—closed to the influx of energy or information—all higher organization eventually crumbles and energy cannot be directed any more to act as power; it all ends up as heat. The conversion efficiency of power plants, which is less than one half of the thermal energy invested, the heat caused by friction in machines or on car tires, or the equalization of temperature between hot and cold water or air in the same recipient, are all manifestations of this law. And since the ultimate system of the universe also used to be assumed as closed, the notion of "thermal death" as the endpoint toward which the evolution of the universe moves incessantly, seemed inevitable. So deeply ingrained became this belief, that the word "entropy" was deliberately chosen as a synonym for "evolution."

Evolutionary views which assume an overall entropic movement not only in the physical but also in the social domain, become fashionable from time to time in materialistic religions. Anarchist romanticists and their followers dream of a pressure-free endstate of maximum entropy in physical and social energy, in which action is neither possible nor necessary in any direction, only a kind of thermal (so-called Brownian) motion of the human "molecules"—

which apparently sounds like Nirvana to some authors (Galtung, 1970).

Other views try to save a static or cyclical cosmos by tacitly assuming an endless world in which entropy would never reach its maximum. Where the old Egyptian and Greek philosophies assumed a cyclical world, Hinduism stipulated cyclical rebirth with an ultimate exit to Nirvana, breaking the monotony of the terrestrial chore. The crudest form of a static, or steady-state, world view seems to have just appeared. This is the ideal of perfect global equilibrium (Meadows, et al., 1972), an equilibrium which is attained only in social and spiritual death, if not in physical death, too.

All theories which have tried to explain the phenomenon of life and its movement toward higher states of organization, and thus at least temporarily toward lower thermodynamic entropy, so far have had to struggle with the difficulties posed by the allegedly unquestionable entropic direction of the physical universe as a whole.[1] The explanations offered were generally of three types:

—*Random fluctuations* are responsible for the *random* formation of macromolecules and forms of life. This explanation, at the core of contemporary reductionism (Monod, 1970), has become extremely unlikely in the light of recent considerations which show, for example, that the number of possibilities for the development of an evolutionary life-chain involving carbon exceeds by many orders of magnitude the number of carbon atoms at disposition on earth; a random process would soon be lost and not get very far in this maze.

—A *deus ex machina*, reigning above the world and separate from it as an independent sovereign, interferes directly and creates some sort of higher accident, unpredictable and arbitrary. In other words, God does not partake in the evolution. Several religions come close to such an explanation—as do several ideologies with their terrestrial gods.

—A special *attractive force toward a finality*, independent of physical energy and partially offsetting the effects of the latter's entropic direction of movement, leads to antientropic "pockets" where higher forms of organization build up at the expense of entropy increase in the rest of the universe. Driesch called this special force "entelechy" and Bergson "élan vital" (vital force). Teilhard de Chardin assumed a psychic (or radial) energy in dialectical battle with physical (or tangential) energy. And Thom (1972) elaborates in topological terms on evolution seen as a conflict, a struggle between two or more attractors, as it had already been stipulated twenty-five hundred years ago by the pre-Socratic thinkers Anaximander and Heraclitus.

---

1. But the concept of thermodynamic entropy is not really applicable to the organization of the human world. In the next chapter I shall introduce more useful concepts.

It is only very recently that some doubts have been cast on the general validity of the Second Law of Thermodynamics. Whereas in our everyday world some of the physical inanimate systems we are dealing with may be assumed to be closed and well in equilibrium, this is not so in an evolutionary world in which galaxies and stars— but also living organisms, social organizations, and spiritual ideas— may by considered as partially open systems in state of nonequilibrium. The new field of nonequilibrium thermodynamics deals with such systems. It has recently discovered the principle of "order through fluctuation" (Glansdorff and Prigogine, 1971; Prigogine et al., 1972): If systems of any kind are in a sufficiently nonequilibrium state, have many degrees of freedom and are partially open to the inflow of energy (information) and/or matter, the ensuing instabilities do not lead to random behavior (even if the initiating fluctuations and the mutation as such are random); instead, they tend to drive the system to a new dynamic regime which corresponds to a new state of complexity. In such a transition, the system acquires new margins to produce entropy, new possibilities for action. A closed equilibrium system, with monotonously increasing entropy, would be characterized by decreasing activity and entropy production. As it gets closer to maximum entropy, which corresponds to the lowest state of order, it would approach an equilibrium state of rest—or death. A partially open nonequilibrium system, in contrast, moves through a sequence of mutatory transitions to new regimes which, in each case, generate the conditions for renewed high entropy production within a new regime, and thus open up the possibility for the continuation of metabolizing activity—for life (see Figure 2). Such systems are characterized by a high degree of energy exchange with the environment and are therefore called *dissipative structures*. Thus, nonequilibrium thermodynamics is leading toward a theory of *self-organization of physical systems*. But, as I shall argue in this book, the principle of "order through fluctuation" seems to form the basis for a new general system theory, valid in all domains: physical, biological, social, and spiritual.

If a large part of the universe may be assumed to be in a state of sufficient nonequilibrium—as, indeed, seems to be the case—we may then come to a revision of the old static cosmos which would be of farthest-reaching consequences: it seems that on the cosmic scale it is no longer necessary to assume *monotonous* entropy increase in all physical systems. Physical energy itself may be an agent in the service of evolution. It would then be superfluous to assume a dualism

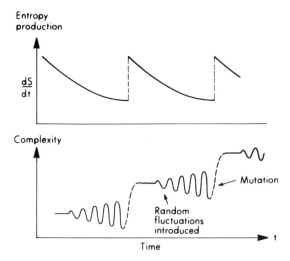

Figure 2. "Order through fluctuation." Sufficiently non-equilibrium systems (dissipative structures) mutate toward new dynamic regimes, which may be at a higher state of complexity, if random fluctuations are introduced. The new regime restores the system's capability for entropy production, which is first high and decreases with rising entropy during each dynamic regime.

between physical and psychic organization—*all organization in the universe would be physical and psychic at the same time.* This is a real breakthrough on the way to a unified evolutionary perspective. For the dualistic view of the Idealists from Plotin to Hegel and of the Gnosis, matter used to be just a waste product of the spiritualization of the universe. But Teilhard de Chardin already assumed a psychic principle—which he later identified as love—to be at work between atoms and molecules in the evolution of higher forms of order in matter. Modern physics is currently looking for "hidden variables" in atoms which transcend randomness and probability and come close to inferring what, in human beings, we would call intelligence. In organic oxidation reactions it has been demonstrated how a dust particle can organize its environment in perfect regularity as if conscious life was expressing itself (the so-called Zhabotinskii reaction). And, at a higher stage of evolution, recent research has revealed psychic phenomena in plants which are comparable to emotions and feelings. Information itself seems to hold self-organizing life and generate more information and thus more order (measured by negentropy). It is only now dawning on us what the core prin-ciple of evolution implies; namely, the self-organization of processes in all domains. In its exploration, the idea of evolution becomes a unifying force of extraordinary power between physical, biological, and social science, the humanities and mythology.

Still, we would need a better understanding of three kinds of evolutionary principles: principles responsible for setting the overall evolutionary direction; principles for the "admission" of individual contributions to the evolution, such as species, and for the selection of "successful" contributions; and principles for the communication between ongoing evolutionary processes and for the linkage with the former two types of principles. Recently developed concepts, which I shall try to outline here briefly, shed new light on these questions.

Older theories—such as Darwin's and Teilhard de Chardin's—assume separate principles for setting the direction and for selecting the participants. Darwin is primarily interested in the latter aspect from which he formulated his principle of the "survival of the fittest"—a principle which is rooted in the environment, not in the process itself. Surprisingly, Teilhard de Chardin eighty years later also assumes the Darwinian principle which is external to the process. Evolution, in this view, would be a broad "trial and error" process which involves the generation of as large a variety as possible of attempts to proceed in multiple directions, only a few of which may be expected to "break through" to higher states of evolution.

If, however, the evolutionary process is understood (as I have discussed it in Chapter 2) as a process generating its own norms, the direction and the selection principles become aspects of the same temporal self-balancing of a process. We may then, with Whyte (1965), distinguish between internal and external factors in evolu-tion, between coordination in the creation of a new form, and com-petition among forms. Whyte's formulation of these factors for biological evolution seems also pertinent in a wider framework, including social, noetic, and spiritual evolution:

—External (adaptive, Darwinian) selection is relative, statistical, deter-mined by comparative fitness, and operates between pairs of popula-tions; its principal criterion is differential reproductive efficiency.
—Internal selection is an intrinsic, usually all-or-none process, operating within single individuals; its principal criterion is the satisfaction of coordinative conditions (whatever they may be).

External and internal factors work hand in hand. But it has to be assumed that internal factors in evolution generally determine first which "evolutionary material" is to be tested by the external factors in the environment. In other words, the overall direction of evolution already expresses itself in the coordinative conditions of internal selection which thus seem to embody the *telos*, or a variational or other finalistic principle, or any other secret which makes evolution take the direction it does. Not randomness, though.

Internal factors in the evolution of forms of life are becoming recognized, or at least speculated upon, at various autocatalytic levels, from polynucleotide polymerization over interacting populations of polymers to dissipative structures at the level of bioorganisms (for a "state-of-the-art" overview see Prigogine et al., 1972). What appears as an internal factor viewed from one level, may be considered as an external factor from the next lower level. Coordination and competition become integral aspects of the same process. This is an important insight also for an evolutionary perspective of man and society: what appears as pluralistic competition at the level of the individual, acts as coordinative principle at the level of human systems. This intricate linkage between horizontal interaction and hierarchical organization expresses perhaps the very core principle of evolution. In biological evolution it is customary to draw the line at the level of the individual, as Whyte does in his above quoted clarification. But if, in this book, we are interested in human individuals as well as human systems, it is important to insist on the complementarity of internal and external factors in evolution. I shall elaborate on this further in Chapters 12 to 14, where this principle will appear as the innermost core of a dynamic hierarchy theory of human systems.

As the apparent dichotomy between internal and external factors becomes dissolved when the self-organizing evolutionary process is considered simultaneously at various hierarchical levels, determinism and finality also become aspects of the same process. The latter aspect is made explicit in a new topological approach by Thom (1972) which stipulates an evolution of forms, a morphogenesis, in all spheres realizing formal structures that prescribe the only possible forms which a dynamics of autoreproduction can present in a given milieu. That such formal structures exist is, of course, in itself a metaphysical assumption. In the biological realm, morphogenesis may be explained both in a deterministic and a finalistic way—both explanations constitute but different aspects of the same evolutionary movement.[1] A local determinism, explains Thom, works through a finite chain of local, relatively well-determined and subtly coupled processes (chreods), which are selected by the principle of minimum production of physical entropy (which gives optimal metabolism) —no equal chance of self-realization—and a global finalism expresses itself in the variational principle of minimum complexity. Both principles may be viewed, figuratively speaking, as "pulling from different ends" and thus eliminating "slack" in the process.[2] The variation of the overall pattern may introduce qualitative discontinuities in the structure of the chreod chain; such discontinuities are then observed as mutations. It is significant that Thom's theory makes instabilities (catastrophes) responsible for mutations. Both aspects, the deterministic and the finalistic, may also be understood as complementary links in a temporal feedback cycle, and Thom believes that metabolism has a weak, but in the long run dominant, impact on the statistics of mutations and thereby shapes the variational principle. Thus, in Thom's topological model—which constitutes the first rigorously formulated monistic model of life— causality and finality become expressions of a pure topological continuity of self-balancing processes, viewed from opposite directions.

This is certainly an oversimplified picture. From a point of view of nonequilibrium thermodynamics, an organism—or any dissipative structure—would, at each transition to a higher state, first increase dissipation and entropy production, perhaps, as Prigogine et al., (1972) suspect, to synthesize the key substances necessary for its survival. Such a behavior has, indeed, been found in chicken eggs and in higher organisms over the first phase of the embryonic life. Only then does the organism adjust its entropy production to a low level compatible with the competitive situation in which it finds itself in the environment. In other words, the phylogenetic interplay

1. As Thom points out, Von Neumann (1963) has shown that also in classical mechanics the evolution of a material system can be described in two ways: (a) by local differential equations, which express the mechanistic aspect of local determinism; and (b) by a global variational principle, which expresses the finality aspect.

2. A similar idea arises from the point of view of hierarchy theory. Levins (1973) and Pattee (1973) suspect partial decomposability to be an objective trait (an internal, coordinative factor) of natural selection in complex systems, and to work toward *self-simplification* of initially chaotic, very complex systems. But instead of depending on unidirectional processes, hierarchical organization may rather—as I suspect—be conceived as emerging from processes running in both directions, "upward" in the organization of systems from simple components, and "downward" through hierarchical constraints in a process of self-simplification.

of internal and external factors in evolution—first the internal coordinative conditions are satisfied, then the external conditions of Darwinian survival—repeats itself in ontogeny; this seems to constitute another profound aspect of the fundamental biogenetic law according to which ontogeny recapitulates phylogeny. But, as Prigogine et al., (1972) remark, it is not at all obvious that evolution can be rigorously characterized by a variational principle, by the monotonous increase of any particular factor. Rather, the factor to be maximized may be expected to be a very complex function of the state of the system. The variational principle itself would then be evolving. Such a stochastic characteristic of finality is to be expected if we assume evolutionary processes of a self-organizing, self-balancing nature.

In Thom's topological model, any geometrical form may be considered information which may be transmitted. The simplest way of transmission, and—as Thom remarks—in the last analysis perhaps the only one, is *resonance* between two symmetrically coupled oscillators, not only in the physical but also in the mental and psychic realms. Quite generally, understanding depends on resonance. Between two static forms no exchange of information is possible in free interaction, but between two metabolic forms free interaction can have large qualitative effects. The topological product of two structurally stable systems is generally not structurally stable; the likely degeneration to a stable field leads to resonance. But it is by no means evident that resonance cannot also occur at a higher level, in particular for nonequilibrium conditions.

From the nonequilibrium thermodynamic point of view, dissipative structures exhibit periodicity and thus are capable of resonance. However, resonance does not only result here in the multiplication of existing forms, but—through the induction of fluctuations which cause the receiving structure to mutate to a new dynamic regime—may trigger true evolutionary processes which lead to higher order and new forms. The thermodynamic approach explains the movement toward higher order in a simple and direct way.

Communication by resonance, which thus seems to be at work in the physical aspects of life, may also be basic for its psychic manifestations and possibly constitute one of the general principles underlying all evolution in the physical, biological, social, and spiritual domains. Diffusion, another communication mechanism, seems to play some role only locally, and not in an evolutionary sense. This corresponds to the general experience in learning and in interpersonal relations where "vibrations" are experienced in many forms, in love, in the charisma acting between people, in sympathy and antipathy, but also in the domain of what is so far referred to as parapsychology. "Vibrations" are also experienced by an increasing number of people as communicating a sense of movement and direction. One "tunes in" to vibrations. In other words, resonance seems also to play a role in the effective communication and enaction of the complex variational principles which guide the evolution toward its finality. In Chapter 9, I shall elaborate on these aspects which transcend our normal rational framework of relating to the world.

Here, I should like to emphasize simply that the emerging concepts of evolution seem to further a view which places all three evolutionary principles—governing direction, selection, and communication—into the process itself. Such a view seems to point to the self-balancing of an overall process which seeks its own norms as it unfolds. In Chapters 6 and 7, where the basic dynamics of the human design process are explored in all their manifestations—art, science, and planning—I shall evoke a principle which is analogous to, or even identical with, this evolutionary self-balancing—and applies even locally so that it becomes observable. In this way, the notion of design may indeed be linked to the notion of evolution, as the title of the book suggests.

Almost all evolutionary concepts have been developed with their focus on biology, sometimes on physics. Social and cultural life is at best referred to in footnotes to such concepts—with one notable exception, Teilhard de Chardin. The exclusion of the human dimension has also blocked a more profound discussion of the meaning of life.

Prigogine and coworkers, on the other hand, *do* restore purpose and meaning—even more: *hope*—to life and human action. If "order through fluctuation" may be assumed at all levels of reality, from inanimate matter to life and to social systems, ideas, and spiritual evolution—then action, the input of energy into dissipative structures, becomes a vehicle for a meaningful evolutionary process. Human design, indeed, becomes an aspect of evolution. But even in this framework of thought the nature of evolutionary finality is left open.

For Teilhard de Chardin,[1] the unfolding process of evolution takes three directions: time, complexity, and man.

A mutation in complexity takes place when a realized structure is no longer capable of coping with the increase in information. In this way, over time, matter developed more complex forms until, over macromolecules, life appeared. The same love which brings people together is at work between atoms and molecules of what we are used to calling inanimate matter. In such a sweeping view, life appears but as a specific manifestation of universal love. Life itself, according to Teilhard, appeared on earth when a stage of complexity was reached in which an excess of information, psyche, remained. But the form life took on earth, the enzymes, cells, organisms, and the whole biosphere we know, represents but one out of many possibilities of life—the one that proved most viable.

Species disappear when they become unfit for life. This stage, Teilhard says, is reached when the original thrust of a qualitative development has gradually run out into a broad quantitative development, into *specialization*. No more need be said to throw a sudden sharp light on the present situation of Western culture.

The process of vitalization led to the appearance of man, to *hominization*—in the light of the most recent paleontological findings, this may have been about three million years ago.

The formation and extinction of cultures is a process in the human sphere which is viewed by Teilhard in analogy to the development of species. Again the same drive toward the realization of new forms, some of which prove viable and form cultures that live for long periods of time, whereas others do not make it. Again the same self-balancing of tentative forms, and again the mystery enshrouding the beginnings of successful cultures that gain life and momentum in apparently a short time.[2] The generation of cultures is part of *noogenesis*, the development of the spiritual sphere, which takes

1. For the following brief discussion of the evolutionary theory of Teilhard de Chardin, I have used writings by Teilhard himself (Teilhard de Chardin, 1959 and 1964), as well as Gosztonyi's excellent summary of all of Teilhard's writings concerning evolution (Gosztonyi, 1968). It should be noted that Teilhard de Chardin formulated most of his essential concepts as early as the late 1930s and 1940s; *L'Apparition de l'Homme* was written 1938–40, *L'Energie Humaine* in 1944. However, as a Jesuit, he was not permitted by the Vatican to publish his work, which came out in book form only after his death in 1955.

2. This, of course, is also consistent with a view of cultures as dissipative structures which organize their internal key elements first before entering a competitive environment.

place at two levels, the individual and the social. Analogous to individual consciousness there is also social consciousness.

Ontogeny, the development of individual organisms; phylogeny, the development of species; and noogenesis, the development of mind, would then be the basic movements toward higher order in the realm of life, working by the same principles. Biological science already knows that ontogeny recapitulates phylogeny. Perhaps the very nature of our *re-ligio* to our own beginnings is to be found in a deep correspondence between ontogeny, phylogeny, and noogenesis. I shall return to this theme in Chapter 9 when I discuss the possibilities of a general genealogical approach to gain evolutionary insight, an approach linking the developments of the body and the mind.

The purpose of noogenesis is the realization of spiritual organization, the *noosphere*, the "consciousness of all consciousness" that, for the Jesuit Teilhard, is centered outside humanity and has a personal center that he calls Point Omega—a personal god. The realization of the noosphere is man's unique role in the world; it is made possible by his capability of mastering matter. Man is a necessary agent for evolutionary communication—for love, not just for life. Although Teilhard is not outspoken in this respect, noogenesis seems to imply the idea of rebirth or reincarnation—or at least some form of evolutionary learning and growth of Mind-at-large through the succession of many generations. Many cultures have adopted such an idea as one of their core beliefs. On the other hand, Teilhard's assumption of a unique role for mankind seems rooted in conventional religious images. Once we take the noosphere to transcend humanity, there is nothing obvious which prevents us from assuming that other phyla in other solar systems, or in other galaxies, are vehicles of the same universal evolution, of the realization of the noosphere.

Teilhard views the overall process of human evolution as a pluralistic process that may be compared to moving along the meridians of a globe, starting from one pole (the common beginning) and leading toward the other pole (the common *telos*, or Point Omega). When this pluralistic movement crosses the equator, it has unfolded itself into the maximum variety and, being "sucked" into its *telos*, starts a phase of *involution*. This stage of crossing the equator is marked by the globalization of the noosphere—in other words, evolution turns into involution in our time. This turn, Teilhard says, is marked by more planning. A balance has to be sought between freedom, regulation, and the overall stream of the evolution. Specifically, the convergence in the material sphere expresses itself through science and technology as well as economic organization; the convergence in the spiritual sphere through a differentiation of the noosphere; and the convergence in the psychic-personal, or social, sphere in the trend toward one humanity.

Calhoun (1971) has supplemented this view on noogenesis in the social sphere. According to his sweeping calculations, man's conceptual space, created by social invention, is capable of accommodating more than a thousand times the number of humans on earth than the physical space alone is capable of—6.3 billion as compared to 4 million. But it is becoming clear not only from these figures, that social space also is now almost filled to the rim, to the last social role.

Calhoun also found a regularity in the process which, over the past forty-two thousand years, or during man's psychosocial evolution, seems to have governed the development of higher social organization, or bigger "social brains," as he calls it. At every doubling of the world population, which came in increasingly short intervals, the old organization form was no longer capable of coping with its problems and its own energy and seven units collapsed into a bigger unit. This succession of collapses, Calhoun holds, may be observed at all levels from families, small tribal groups and full-scale tribes, to nations and power blocks. We are witnessing the end of this development which, of course, implies an alternative between a nonlinear transition toward a new type of development and a rigid endstate. If, however, the process is extrapolated in a straightforward way in order to gain some indication of where we are in it, we would have to expect the last but one collapse around 1985, resulting in the formation of seven large social brains grouping the world population of then 4.5 billion. And world union, according to this train of thought, might then be reached in the first quarter of the next century, at a stabilized population figure of 6.3 billion. (This calculation of Calhoun's takes some nonlinear "flattening" into account). Is this perhaps another indication for "crossing the equator," for starting an involution which will be characterized by processes of a different kind—of global spiritual organization following the completed social organization?

Where, in evolution, do we stand at present? At the globalization of the noosphere, the crossing of the equator, the start of involution, we have heard; at the point where both physical and social space are nearly filled; where the limits to spatial growth impose themselves, we have also heard. Clearly, in this light, we have not only arrived at just another point in time of cultural change but at a major mutation in the evolution of mankind.

Will the beginning of involution mean that less and less new cultural forms will struggle to life, that one principal stem will develop? Has that cross-fertilization of cultures, which was growing in importance during the past millennia, acquired such dimensions that we may soon speak of one ongoing and growing culture? Christian culture all but started from scratch; we are the heirs of the antique world, as well as of others which we do not recognize equally clearly. Already in 1954, Teilhard de Chardin believed he noticed the first signs of a planetary "inter-sympathy" which is created by the search for the same truth, which does not only join our brains, but also our hearts; he felt the "psychic temperature of the earth" rising. Two decades later many more people feel the same. In Chapter 14, I shall try to give some tentative answers to the above questions.

There is no straight prediction to be deduced from a truly evolutionary view which holds that reality is in the event. But it can give us a feeling for the depth of and possible answers to the three questions which resonate in our consciousness: Who am I? Why am I here? Where am I going? It can teach us attitudes and ways of thinking and relating to the world which are more realistic than the causal-mechanistic and morphological (static) modes which are ingrained in Western culture. It can guide us toward "tuning-in" to a reality whose nature is systemic and dynamic, and it can encourage us to accept our predicament as cybernetic actors on earth.

An evolutionary view can also greatly elucidate the tasks, methods, and criteria of planning and design more generally. The task cannot be to predict the future course of evolution and "fit" to it, as an industrial manufacturer may fix his eyes on the anticipated moves of his powerful competitor in order to "fit." The local determinism in life phenomena is of little relevance here. Human ideas and aspirations are of a highly nondeterministic character, although not without some regularity. Evolution is flowing out of the life within us, biological as well as spiritual. In this sense, we are *making the evolution*.

A predominantly Darwinian attitude toward design would emphasize the generation of variety, the creation of a broad spectrum of chances for innovation in the physical, social, and spiritual realms; the basic support of the life process in all its manifestations and the opening-up of channels through which it can flow into the world and achieve a chance for survival in a few viable forms. These principles would then hold for our attitudes toward the lives of individual persons—a friend, a beloved, a companion—as well as toward the lives of social systems and cultures. Not the fixation and realization of targets is in the focus of such design, but the enablement of processes, a kind of enablement which aids them in their natural way of self-balancing and lets them find and build structures instead of imposing structures on processes.

But a crude Darwinism in noogenesis would amount to the implantation of the "law of the jungle" in our technological society. We sense what fatal consequences this might turn out to produce. In fact, no social organization is possible just on the grounds of the survival of the fittest, although some primitive societies may come close to it. A Darwinian principle of selection would correspond to anarchism in all its manifestations—the ultimate in individualism, free enterprise, and nationalism—which Western society has been able to accommodate only partially in the name of liberalism. And yet it is the manifestations of social Darwinism themselves in the form of individual, corporate, and national competition which now endanger the systems of society, the free enterprise of individuals, organizations, technology, information, and ideas. Planning and regulation, and any kind of imposed social order, inherently restrict or counteract the *laissez-faire* of social Darwinism. Design attempts to find, formalize, and bring optimally into play the innate norms of a process. In other words, design focuses on finding and emphasizing internal factors in evolution, on making them conscious and effective.

Thus, the task is to find a way of learning about what Whyte (1965) has labeled the coordinative conditions for internal selection, and what, more precisely, may now be called the realization of innate norms in and through a process. Game theory, within a relatively narrow scope, seems to bring out such innate norms realized in the processes emerging from simple situations of conflict, which may be studied for two or three opponents. Marney and Smith (1971) as well as Laszlo (1972 a, 1973 a) have recently speculated on the innate norms of noetic evolution, of the evolutionary development of knowledge concepts. Other innate norms seem to realize themselves in invention and innovation. It has long been observed that inventions occur in functional clusters and that important inventions have been made twice or more often in the evolution of mankind. Agriculture had been invented about twelve thousand years ago in the Middle East and, without apparent connection, six or seven thousand years later again in America. The idea of a "periodic table of technologies" (Ozbekhan, as quoted in Jantsch, 1972) which, in analogy to the Mendeleyev system of chemical elements, would connect essential features of potential technologies, has not made much progress; it would permit the recognition of some of the innate norms which obviously act in technological progress—which, in our day, has almost assumed self-organizing and self-regulating characteristics —and which may be linked back not only to norms innate in technological development but also to human inventiveness and feedback interaction with the environment. Gabor (1970) has recently pointed to the mismatch in the norms realizing themselves in technological and social developments, since the former have assumed semiautonomous characteristics.

There is nothing firm for us to grasp. We have little experience yet to draw from and no theory of social involution. All we can anticipate is that constraints will increase in physical as well as social space and that the freedom which evolutionary impulses used to enjoy in living themselves out will look different in more "crowded" conditions. We can also anticipate that these conditions will emphasize the systemic interaction between many forces and processes—which seems to imply that regulation of our own and our systems' movements will become more necessary and more difficult than ever. It is not yet clear precisely what is the role of this regulation of human affairs against a global evolutionary background—nor is it clear whether regulation plays different roles in different domains of human life. But though we do not see the stream, we must certainly strive toward sensing it. We have to develop our physical, social, and spiritual—above all the spiritual—sensorium for the stream of evolution. This is a difficult and, in global dimensions, a novel task as well —but not a hopeless one.

In Parts III to V of this book I shall discuss what can be said about these questions at this stage. But before dealing with the human design process I shall explore in the next chapter the nature and ways of self-organization and functioning of human systems, which form the spearhead of terrestrial evolution in its present phase, the highest form of organization noogenesis has so far brought forth on the earth.

HUMAN SYSTEMS AND
SELF-ORGANIZATION

A broad evolutionary view provides us with a basis of principles which are at work at *all* evolutionary levels, including the level of the systems in which and through which human life evolves and organizes itself. This important aspect has been generally emphasized by evolutionary thinkers through the ages. The dynamic situations governing the evolution of form in the physical and biological domains are basically the same which govern the evolution of man, society, and culture. Plato saw in them the very justification of human systems, which he gave magnificent expression to in his *Timaios*: "There is only one way in which one being can serve another, and this is by giving him his proper nourishment and motion; and the motions that are akin to the divine principle within us are the thoughts and revolutions of the universe." This identification has important consequences in several respects. For example, it justifies an anthropomorphic view of evolution and the use of anthropomorphic metaphors in the description of any aspect of dynamic reality, including even physical science.

Thus, an evolutionary monistic view may regard all human systems —organizations, institutions, cultures, and so forth—as having life in a real sense, life which in its basic processes and manifestations does not essentially differ from the life which we recognize in individuals and which has not only a physical and a biological, but also a social and a spiritual aspect. The model of the basic human design process, which I shall develop in the next part of the book, tries to apply these common basic principles.

Man lives in physical space like other creatures—but not only in physical space. When he emancipated himself from nature and embarked on his psychosocial evolution—which, we are told, has now been underway since the Neolithic age, some forty to fifty thousand years ago—he started to build for himself a conceptual space which is the realm of his mind as well as his feelings, of his imagination and understanding, perception and conception, of his own will and of his creative urge and capability. Man is the only animal on earth to have fully developed such a conceptual space.

The conceptual space has at least a social and a psychic or spiritual aspect; it is perhaps sufficient for a first approximation to distinguish between just these two. Social space is created through differentiation, through the design of social roles and systems of such roles as they come into being with all kinds of social invention and innovation (Calhoun, 1971). Labor and task-sharing, crafts and industries, trade and the accumulation of wealth, government and education, physical and conceptual mobility, collective health and organized fighting—they are all social inventions with their corresponding patterns and systems of roles. Spiritual space holds man's relations with the numinous; his quest for purpose, direction, and meaning; his cultural inventions from values to religions, from the arts to philosophy and science.

If human systems are called those systems in which and through which human life evolves and organizes itself, what precisely is to be understood by this term? Simply social systems of various size and scope? They are justifiably moving into sharp focus today when we speak of design, normative planning, political processes, conflict, and the like. But men also live in a physical and a spiritual world. Obviously all these worlds interact in forming what we may perceive as a social system. Physical systems, such as the buildings of a city, its networks of streets and of telephone connections, shape the life of the community living in the city no less than the image and the values its inhabitants attach to the city. Bertalanffy (1967) points to the reality, in an operational sense, of values and ethics in the life of complex human systems.

In Figure 3 I attempt to sketch the processes which structure, and in turn are structured by, human systems. Human systems may then be discussed in terms of six basic aspects corresponding to these processes. They form a self-balancing loop of the general kind I have postulated in Chapter 2.

On the one hand, Figure 3 organizes the aspects of human systems which correspond to the three spaces in which man lives simultaneously: physical, social, and spiritual space. These three spaces also underlie the structure of the human design process as I shall develop

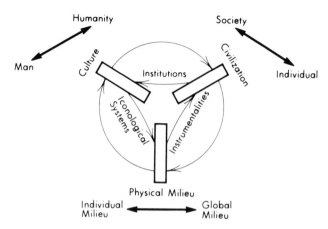

Figure 3. The process structure of human systems. The transformer systems (iconological systems, instrumentalities, and institutions) constitute interacting feedback processes which give rise to the "standing-wave patterns" of the interference systems (physical milieu, civilization, and culture).

it in Chapters 6 and 7. Here, I shall assign to them the corresponding notions of human systems aspects which I call *interference systems* because they are generated by interference of many processes:

—The *physical milieu* holds the totality of *physical systems* in and with which man interacts. Examples are man/machine systems such as a person driving a car; whole networks such as road systems, telephone networks, railroads, water distribution systems, and power grids; natural ecological systems; but also groups of people physically inter-acting with each other, such as work teams or military units.

—*Civilization,* in the meaning given here, comprises the totality of *social systems* with their multiplicity of interactions between their members, forming a "tissue of reliable relationships" (Vickers, 1973). Families of various kinds (marriage, communes, etc.) and communities from the local to the global level would be examples. Their structures are less significant than the types of interaction that characterize them, in particular interaction by means of economic and political processes involving persons as well as technologies that form part of the system. To emphasize such aspects one often speaks of socioeconomic, socio-political, or sociotechnological systems, which sounds really misleading if these terms are understood as pointing to an internal juxtaposition between social and economic, social and political, or social and technological aspects as if these were "solid" systems elements, not processes.

—*Culture,* as a minimal notion, comprises all aspects of a shared appreciative system—"carved out by our interests, structured by our expectations, and evaluated by our standards of judgment" (Vickers, 1970)—plus a communication system through which sharing becomes possible. Its core is a system of shared values.

These three aspects of human systems are far less tangible than they may appear at first sight. Therefore, I have drawn them as thin slabs, or membranes in Figure 3. They do not represent anything approaching a solid and extended entity, with inflows and outflows through which it may gradually change. Rather, they may be viewed as a kind of standing-wave pattern shaped by the interplay of various processes and subject to mutations in accordance with changes in the system of processes. It is perhaps not exaggerated to expect the future development of some kind of social and cultural wave-mechanics determining the eigenvalues of human systems; with such a science we may become able to determine criteria and thresholds for mutations in human systems.

A process-oriented view has also been made the conscious basis of recent experiments in building communes as feedback learning experiences relating to complex human ecosystems.[1] Even physical systems may be viewed as manifestations of interactions rather than as entities in their own right. Recent examples have become known in the United States where the physical construction of hospitals has commenced with no fixed plan in mind, but with an interaction of processes involving all interested sides shaping the functions and eventually the physical structures as construction work proceeded.

The above listed human systems aspects, which I called inter-ference systems, may be regarded as interfaces between agents of another kind which organize the processes I mentioned, and which may be named *transformer systems*. I am speaking of the carriers of the feedback processes that appear in Figure 3 as links between the physical milieu, civilization, and culture. The three types of trans-former systems are the following:

—*Instrumentalities* link the physical milieu with civilization, transforming both through their activity. Examples are human organizations such as specific governments at all jurisdictional levels, corporations, uni-versities, citizen groups, or judicial courts, along with their tech-

---

1. Perhaps the most significant experiment of this kind is "Project 1" in San Francisco, which started in 1971 with the physical and social transformation of the amorphous space of a big warehouse into a complex and subdivided living space, a community of about one hundred people, with the idea in mind to learn from the interaction of processes for later application to social eco-systems of larger size (perhaps up to three thousand people).

nological means and communication systems. The role-playing pattern of instrumentalities and their individual members build a civilization by transforming the physical world and molding it together with people into social systems; in turn, it also transforms civilization by fitting it to the physical milieu—as we so strongly experience today when our lives are increasingly ruled by technology.

—*Institutions* mediate between civilization and culture, introducing specific sets of values into the framework of social systems—staking a claim for them, so to speak—and building cultures through their role-playing patterns which shape the value systems at the core of a culture. Examples of institutions are government in general, business, education, churches, law, public health, consumer protection, and so forth. Institutions are usually the general expression of values which are acted out in the particular form of various instrumentalities.

—*Iconological systems*, finally, provide the important, though not always explicitly recognized, link between culture and the physical milieu. They mediate between symbols and their interpretation, between form and content, between value and physical manifestation. They structure the processes which create and maintain man's cultural world "by the symbolic transformation of the actual world and the imputation or projection thereon of the meaning and values by and for which we live" (L.K. Frank). The most important examples are mythological and scientific systems, the former emphasizing subjective participation in a holistically viewed universe, and the latter objectivity and separation. Language and other systems of communicability, as well as artistic forms of expression, belong here and may be understood in terms of mythological or scientific systems, or both.

To represent transformer systems, such as instrumentalities and institutions, as processes rather than structures may look odd when one thinks of organizations with well-defined administrative or work structures, such as business corporations or schools or bureaucratic agencies. Yet the processes again are what is decisive here. If there is much emphasis on structure, this means that only a particular inflexible pattern of processes is allowed to play. On the other hand, the need for more flexible instrumentalities has recently led to the conception of instrumentalities built of interactive processes with an absolute minimum of structure to guide them.[1]

In Figure 3 I also tried to make visible as a third dimension the spectrum between single man and the totality of mankind which

1. The new Medical School of the University of California in Berkeley, operative since 1972, is a good case in point: it has no buildings, no clinic, no administration, and no faculty of its own. It is defined by the interaction of faculty from various departments of the Berkeley campus, and from outside (including doctors in local hospitals who also "bring" their facilities) with the students. An interdisciplinary "Design Group" is planned in Berkeley along the same principles.

marks human systems. While writing this, I am forming a human system with the table at which I am sitting, the typewriter which I am using, the lamp and its lampshade with Indian miniatures which stimulate my mind and sensitivity, and the lush green of the mother-of-thousands plant which brings life to my working place. Couples, families, tribes, small and large organizations, cities, states, and armies with all their respective technologies form human systems. A human system may have any number of people as members, up to society-at-large and mankind, and the global milieu of Spaceship Earth and the "global village." There are institutions which number only one member, such as the president of the United States, whereas others, such as business, have many millions of members. A culture may have no bigger base than a tribe or a sect, or it may approach global dimensions.

A fourth dimension, not indicated in Figure 3, is time. All these systems change and evolve. They are never for long in a stable equilibrium. At best, they attain a precarious and temporary equilibrium when a culture peaks and exerts a strong stabilizing influence on the other human systems aspects linked to it. The normal case is that of a mutation rocking any part of the overall human system— a culture if new values come into focus, or the physical milieu as we experience it now with the degradation of the environment, or a civilization if it becomes unmanageable without reordering the whole human system loop—and the subtle play of the total loop seeking to balance out the ensuing instabilities by adjusting all aspects of the human system.

The six basic aspects, or subsystems, of human systems are not to be understood as entities that exist side by side with different memberships. On the contrary, any individual person is typically a member of human systems in all its basic aspects and generally adheres to more than one manifestation of each aspect. We are all multiple members of various institutions. It is this extension of the individual over the whole loop which gives human systems their identity and life. Each individual strives to regulate his own life so as to reconcile the often conflicting demands emanating from his membership in various subsystems; for Vickers (1973), this precarious struggle even constitutes "the basic art of being human." A young professor in Berkeley, in order to win tenure, needs both popularity with the students and the benevolence of his older colleagues who determine the fate of his department. Therefore, he will wear a beard and long hair and talk moderately radical in his lectures but, at the same time, comply with the absurdities of the grading system and write his papers in terms of behavioral science or any narrow model prescribed by his department as the only valid one in scholarly discourse. If he is honest with himself, he may not be able to live with the conflict imposed upon him by the demands made from both sides. If he can bear it, and without giving in, he may bring new life to the institutions and instrumentalities whose membership are causing him his headache, and he may even become a rebel overthrowing an iconological system in his scientific discipline.

A hierarchical systems representation is not capable of fully expressing this intricately interwoven play of interests and acts by individuals. Neither a rigid hierarchical system controlled from the top, nor a multiechelon system in which the goals build up from the bottom level and are coordinated from the next higher level, can be a fully valid model here. Such representations assume that holistic entities—people, units of people, organizations—can express themselves unambiguously and in a total way on any issue, so that they can be represented in a block diagram as distinct elements of a system. In reality, we find individual man in his search for his identity, his oneness, participating in many ways in the subsystems of all the above-outlined basic types, radiating energy in many directions, and drawing from many of them. A worker cannot be viewed just as a human element contributing to the production processes in a manufacturing company; he is probably at the same time a member of a union and may, in times of conflict, act against the interests of the company as defined by its management; he may also be head of a family and therefore vote for ending a strike called by the union before its targets are attained; and so forth.

Human systems thus are typically expressions of specific aspects of human nature, of particular features of humanness manifest in its personal members. The organismic image of human systems stipulating a hierarchical relationship between members and systems in analogy to the hierarchical relationship between cells, organs, and living organisms—as it appears in the light of current conventional wisdom—can only express a partial aspect and in any case is grossly oversimplifying. Cells are not multiple members of different organs. But participation by each individual in the entire loop of the six basic aspects of human systems ensures close coupling of the forces acting in the self-balancing evolutionary process. Multiple participation of individuals in subsystems of the same basic type ensures tension and movement in the latter. Multiple demands on individuals from subsystems of the same type ensure tension and movement in the individual. As individuals impart life to human systems, the latter, in turn, stimulate life in individuals.

At the same time, the type of free interaction between metabolic systems which constitutes the principal or even sole process of communication among biological systems, forms only one—and perhaps not always the most prominent—process of communication within and between human systems. In certain situations, the joint membership of persons in several, even conflicting, subsystems may prove to be more important. It gives rise to complex interactive patterns of individual role playing. A suitable image would be that of a multidimensional, intricately interwoven web of relations, activated and continuously changed through all the individual "nodes" they are bound together with and from which they radiate. If one particular relation in this web becomes activated, both "nodes" (individuals), between which it extends, change their vibrations and perhaps get into resonance with each other—but, at the same time resonance phenomena may occur in any part throughout the entire web (the entire human system).

Interestingly, Thayer (1972) suspects that it is excommunication, the blocking of communication, and thus the creation of nonequilibrium, that leads to mutations and the emergence of new forms in open communication systems—again, "order through fluctuation," where the fluctuation is provided by an entropic move.

As I shall outline below, much speaks for the assumption that resonance constitutes the basic mode of communication in human systems, as well as for many other types of system. Resonance also seems to underlie the notion of individual participation in a supraindividual group consciousness, as stipulated by modern psychological theories (Gosztonyi, 1972), and may play a still bigger role at all systems levels from the family to mankind. And human systems themselves may participate in a wider, universal consciousness. In Chapter 9 I shall briefly introduce some of these concepts, but we do not yet know much about them. In any case, we may assume that communication within and between human systems works at both the personal and the systems level, bringing into play a rich variety of potentials. Later in this book I shall review the nature of the communication processes in human systems from various points of view.

Laszlo (1972 b) suggests that all natural systems, from atoms to societies, possess subjectivity, or nonreflective consciousness, though of different grades. Human systems certainly have purposes and goals, as individuals do; they have consciousness with which they can recognize and honor the systemic and dynamic nature of all life processes. And they may even be capable of reflection—or how else would the Zeitgeist, would social values and whole cultures come into being?

In short, with an evolutionary view as a basic assumption, social and cultural organization finally become *theoretically* possible—where, indeed, they have proven *practically* possible and even thriving in so many variegated forms during mankind's psychosocial evolution. It is not out of pure irony that I state this. I believe that an evolutionary view is the most fruitful paradigm, if not a necessary prerequisite, for rewriting social science as a science of human action instead of a science of mere behavior. If reality is in the act, not in the setting in which it occurs, action is inseparable from the moral responsibility which it carries.

Human systems also distinguish themselves in another important way from biological systems, namely through their capability of generating and bringing into play information almost instantly, or at least in negligible time spans compared with the mutations which evolution introduces in biological systems. We may say that human systems are capable of invention. Cognition, insight, ideas, models, expectations: all introduce what, in biological terms, would be called extremely unstable chreods (development lines) which interact and compete with each other. This specifically human capability of invention accelerates the evolutionary processes to such an extent that they become directly observable in the form of history and even in the form of cultural mutations. As Calhoun (1970) puts it, the deviance controlling processes is always slightly subordinate to the deviance promoting processes—"man is a crisis provoker."

The mechanism through which invention becomes effective in human systems is still puzzling social scientists. This is largely their own fault since, in their quest for "scientific" rigor, they have never attempted anything other than turning to the models of physical science. It was perhaps the future-orientation of physicists, their concern with objective predictability of physical phenomena, which drove social science toward analogies rooted in mechanics rather than in biology. Thus social science became an expression of an equilibrium steady-state world, of a static Greek cosmos incapable of accommodating the evolutionary dynamics of cultural change, which is, therefore, simply ignored or even denied by traditional social science.

Social science has imitated the models of Newtonian mechanics as well as some of the more refined models of statistical mechanics. Newtonian mechanics of strict cause/effect relationships still governs much of social theory. It underlies the "social physics" models of history, such as Marxism with its laws of motion for capitalism, the ends/means approaches to social policy and much of economics. More fundamentally, it serves as a justification for applying technological modes of planning and action to the social fields, in particular sequential problem solving—as if there were in society, like in technology, problems which could be solved once and for all—and priority-setting. The entire school of social science regarding social systems primarily as information-processing devices, devoid of any life of their own, is under the spell of Newtonian mechanics; some versions of the policy sciences are included in this class.

A step further, social science came to imitate statistical mechanics and marry it with behavioral concepts. The resulting model, which has become basic to social science in the mood of applied behavioral science—and almost obligatory today in Anglo-American academic environments—resembles a "kinetic gas theory of society" which is as ugly a concept as it sounds. It assumes that the behavior of groups can be predicted with statistical accuracy no matter how far individuals may deviate from the average. Statistics has become "the inductive technology of social science" (Rapoport, 1970). Even if individual deviant behavior is taken into account in a predictive model in the form of differing responses to signals in an evolving situation, this is still done on a statistical basis using information collected in the past and thus locking the predicted future into configurations of the past. One of the foremost tasks of our time is the rewriting of social science as a science of human action, of design—where action and design are understood as irreducible to behavior.

But physical analogies may nevertheless become useful in social science if they match to some degree the conditions prevailing in human systems. The model of statistical mechanics which was conceived at the end of the nineteenth century, deals with closed systems near equilibrium and assumes collision times that are very short compared with the time spans between collisions. Such brief, sharp interactions are not characteristic of human systems. Rather, we are dealing with near-continuous highly nonlinear interactions between and within finite systems which are partially open to new information, and thus to an inflow of energy, and which rarely come close to equilibrium conditions. It seems that the new and exciting field of nonequilibrium thermodynamics, I mentioned in the last chapter, may provide useful analogies. It assumes finite, partially open systems, collision times of finite duration, and memories. Such systems, characterized by a high number of degrees of freedom, exhibit dynamic behavior less random than that of equilibrium systems (Glansdorff and Prigogine, 1971). Recently obtained results show that under such conditions the product of the interaction between two highly nonequilibrium structures (such as strong flows) may be of highly regular organization. The resulting dissipative structures, of which life is a general example, permit periodicity which, in turn, brings in resonance as a general mode of communication.

The theory of dissipative structures, first conceived in the realm of physics, seems to transcend all boundaries set by the nature of systems and, indeed, constitute a general theory of the dynamics of systems—in the physical as well as in the nonphysical realm, for animate as well as inanimate matter, for matter as well as for spirit. This is a hypothesis at present, but one which seems to hold the central paradigm that permits the extension of the concept of evolution over all aspects of reality, including humanity and its systems. In a future physical analogue for social science, human systems with all their tangible and intangible aspects might then perhaps be regarded as dissipative structures, arising from the interaction of strong and highly nonequilibrium flows of ideas and actions. Their spatial organization would then be the result of processes of self-organization and free interaction, as would be their temporal organization, or in other words the forms of periodicity built into human systems. This organization would be physical as well as psychic. Indeed, the borderline between both becomes blurred in the light of the emerging insight that information itself may have a self-organizing capacity, that a seed of information may engender more information and thus more order (Tobias, 1973).

I am not suggesting here any reductionism of human systems in terms of individual beliefs, ideas, and actions. Rather, I am suggesting an evolutionary process within human systems whose fine-structure shows a multiplicity of ideas, initiatives, and relations struggling to life and their selection being governed by an evolutionary law effective in the system in question as a variational principle or in any other form. In times of a widely shared cultural basic, such an evolutionary principle acts in a highly selective, almost deterministic way, whereas in times of cultural flux there is a confusing pluralism of noetic life advancing on a broad front. As humans shape the systems in and through which they live, they are in turn shaped by their human systems.

Understanding human systems in terms of dissipative structures would provide a theoretical basis not only for social and cultural organization (as role playing and multiple membership also do) but for *self-organization* toward higher states of systems organization. Such a theory would also deal with the effects of introducing, or adding, instability to human systems. It may become capable of explaining cultural mutations that so far have seemed to defy any explanation in terms of energy flow (or the exertion of power). Weak and defeated countries often play out their culture as an element of instability to which the powerful and victorious countries become exposed—with the result of a mutation in the latter. A nonequilibrium social theory (including economic and political aspects) would also find an extremely fruitful field of study in the question of what type of instability ought to be introduced deliberately, and when. The whole value structure embedded in conventional social science and emphasizing stability would become reversed in a new "nonequilibrium social science" geared to the understanding and furthering of self-organization in an evolutionary perspective!

Such a perspective would come natural to human systems which even powerfully enforced religions or ideological models cannot close off for too long a period. In a human system, individual membership changes because people die, and new aspirations and ideas enter continuously from within the system. Human systems have memories which hold not only images from the past but also images of the future which introduce the most powerful fluctuations upsetting any mechanistic stability. These images of the future play an important role in a type of autocatalytic process I shall develop in the following chapters and shall call the *human design process*. But human systems are quite generally characterized by a large variety of autocatalytic processes, or positive feedback loops, bearing on economic as well as on physical, social, and cultural aspects. Stability is not a notion which might govern human systems over very long periods. This insight is just another expression of the general recognition that systems of high complexity trade stability for richness in adaptability (Laszlo, 1972 a). The strong coupling between subsystems (in particular through lively communication) leads to conditions for metastability which become more favorable with higher complexity; human systems may be suspected to be generally in a metastable state (Prigogine, 1974 b) and thus "ready" for mutation once the fluctuations attain critical size.

In fact, the Prigogine principle of "order through fluctuation" may well be empirically discovered in many well-researched aspects of human system dynamics. The basic characteristics are the same as in the physical realm: A human system, stifled by high entropy, absorbs energy or information inputs (fluctuations) which cause it to fluctuate and eventually mutate to a new dynamic regime. In the new regime, entropy production is first very high due to internal coordination with the aim of forming a new structure corresponding to the dynamic regime, and subsequently becomes increasingly governed by economic criteria—minimum entropy production in an external, Darwinian process of competing with other systems in a shared environment. The graphic representation in Figure 2 (see p. 38) holds in the same form for physical and human systems. To illustrate this, I shall consider three examples:

If a city's traffic arteries become clogged, a new freeway introduces a new dynamic regime. When the new freeway opens, much entropy is produced by the restructuring of communication processes, with more people driving into town than before and new patterns of business, service, and leisure activities developing. In not too long a time the new regime will have run into high entropy again and become clogged.

As a second example, I imagine an office in government or business which operates in a dull, bureaucratic spirit, which is the same as saying it operates in a state of high entropy. Into this office breaks a newcomer, a bright and ambitious young man, straight from the university where he has toyed with many ideas he now wants to test in real life. If the office is an equilibrium system, with everybody sharing the wish to operate in a minimal way and to resist change, the young man will be recognized as an intruder and become encapsulated, or transferred at the first occasion. If, however, the situation is of a sufficiently nonequilibrium character, with tensions and dissatisfactions smoldering under the surface, the arrival of the young man may easily trigger a jump to a new and more ambitious form of life and work in the office. After an initial exciting period of innovative ideas and the formation of synergistic links among people, things may settle down again to a new, increasingly dull routine—until the next intruder arrives.

Finally, let us consider a revolution in a country in which the hard grip of power had been felt. Entropy is usually high in a dictatorial political system—it is not the degree of "law and order" which is essential here (a high degree of "law and order" would then suggest low entropy), but the degree to which the creativity of its individual members can unfold and contribute freely to the life of the system. A dictatorial regime is always very narrow in the contributions it is ready to absorb. Thus, entropy is high and the blocked creativity, in the form of rebellious or merely neurotic behavior, introduces fluctuations which eventually lead to a mutation, a revolution. Never is the world so beautiful as on the first morning after a revolution. Everything seems possible, and "everybody" is called upon to contribute to building a new human system of creative processes, with a new dynamic regime as unstructured and open as possible. But much of the initial enthusiasm is subsequently channeled into building specific structures, power builds up in a new configuration, the options and the welcome forms of creative contribution are narrowed down, and entropy builds up again.

This last example suggests an interesting insight into the role of complexity in human systems. If order is considered to be equivalent to the amount of information required to describe it, a mutation in a human system increases order, or complexity, whereas power tends to decrease it. Quite generally, we may recognize here a *basic equivalence of power and entropy*, which may also be translated into an equivalence of rigid structure and entropy. In this book I shall repeatedly return to the emerging guiding principle of a new approach to the design of human systems, focusing on maximum flexibility and, to the extent possible, on freely interacting processes instead of structure. Such an approach facilitates the self-organization of higher complexity and thus a higher form of life for the system in question. It gives the evolutionary mechanism of "order through fluctuation" an optimal chance. The Maoist ideal of a "continuous revolution" finds at least a meaningful theoretical basis in general system theory.

It also becomes clear from these examples that internal and external factors of evolution form a peculiar bond in the hierarchy of human systems from individuals to social systems, cultures, and beyond. They are different aspects of the same process. What appears as competition, or external (Darwinian) selection, at the level of individuals, becomes coordination, or internal selection, at the next higher level of a social system. External interaction among individuals lays the internal foundation for a new system regime. In Chapter 14 I shall return to this principle and formulate it in a more general way, Here, I wish merely to point out that "order through fluctuation" seems to be a basic mechanism penetrating all hierarchical levels of human systems and responsible for mutations in organizations, institutions, and cultures as well as in the overall dynamic regimes of mankind at large, which evolved from hunting and fruit collecting to primitive agriculture all the way to our current global systems of cooperation. Social science has so far generally recognized external, Darwinian, factors only. In this book, I shall try to shed also some light on internal factors in the evolution of human systems.

Technology tremendously reinforces many of the individual development trajectories, or chreods, as Waddington (1970) calls them, but also tends to stabilize and rigidify them. Man, in his physical organization, is a multifunctional being to an extent surpassing all other forms of life. Where animals survive on the strength of one or few senses, functional capabilities, and possibilities for adaptation, man has a large number of them at his disposition to bring into play in a flexible and combinatory way. These functional chreods are separated from each other and reinforced, first by simple specific tools, then more generally by technology which thus becomes the major vehicle for specialization and social differentiation. The same applies to the social chreods which become manifest in roles and role patterns at the individual, instrumental, and institutional levels. Man's multiple roles become strengthened and, at the same time, resistant to change through specialization and task differentiation.

Technologies can become the core of major mutations in human systems. But the wrong design for bringing them into play may lead to disastrous results. Hauser (1973) compared the sequence of technological revolutions which gave rise to a thousand years of Western success in economic and social development to the different sequence of the same technological revolutions imported to, or imposed upon, the Third World in the "wrong order." Mutations in the Western world built upon each other in the following way:

—A *commercial revolution*, in the twelfth century, created money and differentiation in trade and produced an affluent middle class in between feudal barons and farmers as well as a capital basis for the following two mutations.

—A *revolution in war technology*, in the fifteenth century, led to greatly increased options in transportation (oceans).

—The *agrarian revolution*, around 1700, improved food production in a quantitative and qualitative (animal protein) way; higher productivity in food production also liberated the necessary labor force for the following mutation and further developed the capital basis.

—The *industrial revolution*, usually dated at 1786 (mechanical loom), was able to build on the prerequisites of adequate food, labor, and capital bases.

—The *medical revolution*, which may be dated at 1796 (smallpox vaccination), led to a population explosion which found adequate food, work, and (at that time) space, including the colonies, to unfold without much difficulty.

—The *transport and communication revolution*, in the middle of the nineteenth century, which made wider sharing in material well-being possible and, in a compounded effect, saw a gradual stabilization in population figures.

The sequence of the same technological revolutions, introduced by the Western world to the Third World in ways which served Western interests best, looks different:

—A *revolution in war technology* cemented feudal rule and was even meant for that.

—A *commercial revolution* was used by the ruling class to orient production toward Western needs, which led to the impoverishment of local crafts, to higher exploitation of agrarian income, and generally to a sharper economic and social stratification.

—A *transport and communication revolution*, Western style, as the ruling class welcomed it, was possible only through more exploitation, which in turn aggravated economic and social stratification.

—A *medical revolution* led to a population explosion which did not find adequate bases of available food and work.

—An *industrial revolution*, Western style, emphasized capital-intensive, labor-saving production methods under conditions of abundance of labor and lack of capital.

—An *agrarian ("green") revolution*, starting only in the second half of the 1960s, came much too late and may now encounter ecological constraints generated by the population explosion.

The importance of design in regulating the conditions for mutations cannot be illustrated more dramatically. What used to grow slowly and "naturally," has now become a matter of pointed action —a matter for design.

With technology reinforcing specific development lines, it is not surprising that in the human systems loop according to Figure 3 it is apparently the transformer systems—instrumentalities, institutions, iconological systems—which tend to rigidify first and stifle the balance of the whole loop. In our time we have become painfully aware of the necessity of institutional change. Yet, a "business as usual" attitude prevails and many people find security in the continuity of an established framework of institutions. In such a state of rigidification the transformer systems promote quantitative rather than qualitative change, which then may degenerate to the kind of monstrous growth syndrome threatening to bury us. But such a state, as I pointed out in the last chapter, also seems to signal an impending major mutation in evolution. Such a mutation, it seems, may well appear first through change in the iconological systems, in particular in the mythological relations linking man to reality.

For the purpose of characterizing the self-organizing behavior of human systems, we may look at the systems from two different points of view: how flexible they are in changing their internal structure—this aspect may be called the internal self-organizing behavior of systems; and how flexible they are in changing their pattern of interaction with the environment—the external self-organizing behavior of systems.

We may distinguish between the following three basic types of *internal self-organizing behavior*:

—*Mechanistic* systems do not change their internal organization;
—*Adaptive* (or organismic) systems adapt to changes in the environment through changes in their internal structure in accordance with pre-programmed information (e.g., engineered or genetic templates);
—*Inventive* (or human action) systems change their structure through internal generation of information (invention) in accordance with their intentions to change the environment.

In mechanistic and adaptive systems, information concerning their internal organization is preprogrammed, e.g., through engineering or genetic/physiological control. In inventive systems, such information is generated within the system and in feedback interaction with the environment. This, at least, is the picture we gain if we apply our human time scale in which genetic evolution acts very slowly compared to human intelligence. But we may also say that adaptive and inventive systems both play their roles in the unfolding of evolution in their own way, and at the scale of their proper time, the time which they carry in themselves. The evolutionary time scale for adaptive systems in the biological domain would then correspond to the unfolding of biogenesis, whereas for inventive systems in the human domain it would correspond to noogenesis. From our human, noogenetic point of view, biogenesis appears to be acting very slowly, but in its proper domain it works by the same principle of internal generation of information, or invention. The distinction between adaptive biological and inventive human systems in an evolutionary perspective, therefore, is mainly one of external time scale applied. But in a cosmos creating itself, time is not an *a priori* category, but unfolds with morphogenesis, with the unfolding of form.

Form and time, then, unfold at many different levels, the interplay constituting the process of evolution, which cannot be grasped through an absolute and uniform time scale, just as it cannot be grasped through any absolute spatial measure. Physical mechanistic systems, in such an evolutionary perspective, would then perhaps also exhibit inventive features, but at still another, slower time scale than bioorganisms. The self-organization of dissipative structures is not restricted to the realm of life but is a general characteristic of all evolving form in the universe. Teilhard de Chardin's idea of love between atoms and molecules, mentioned in the last chapter, is not as bizarre as it may seem at first look—it is about to find its scientific formulation.

In many cases, the theoretical treatment of mechanistic systems is possible through linear systems of equations. Most engineering systems are therefore built as mechanistic systems, but also the bulk of econometric approaches to forecasting may be considered as mechanistic—and thus as unrealistic if applied to any social system not controlled in a centrally planned way. Adaptive systems, representing "feedback through parameters," already require complex nonlinear forms of negative feedback approaches for their theoretical consideration, amenable primarily to computer simulation. Inventive systems would in addition require theoretical approaches dealing with the introduction of complex positive feedback elements into dynamic situations to gain some basic understanding of their strong interaction with the environment.

Wherever social systems aim at changing their own internal organization, the political processes for planning and enacting such changes become an integral aspect of their regulation. Only for rigid and centrally planned social systems under stagnating modes of control do political considerations remain outside the planning area—they are prefigured already.

Inventive systems, beyond internal political processes, also have to take into account ecological feedback relationships between the system and its environment. Since the environment is partly shaped by the system to which it reacts, the external behavior of the system also ought to become part of the political structure; recognition of this is only gradually emerging through the general debate around the deterioration of the environment due to man's interventions. But the evolutionary and anthropological aspects of this feedback interaction belong partly to cultural change beyond the grasp of conventional political processes.

The classification of systems by internal self-organizing behavior points to a corresponding stratification of levels of change. Expressing change in Aristotle's terms, locomotion or quantitative change would characterize a mechanistic system, qualitative change an adaptive system, and generation/corruption an inventive system. In modern terminology, we may distinguish three corresponding modes of social change: force, role playing, and syntony. In the following chapter, it will also become clear that change is viewed quite differently at three different levels of inquiry. Perception, change, and the self-organization of human systems seem to be structured by the same hierarchy of levels—as, indeed, seems to follow if in an evolutionary perspective it is the event which is set absolute.

External self-organizing behavior of systems, i.e., their way of interacting with and acting upon the environment, may best be discussed in terms of the basic three-level structure of planning proposed by Ozbekhan (1969) which I have further elaborated (Jantsch, 1972) and to which I shall return in Chapter 12: (a) normative or policy planning—focusing on the enduring themes of dynamic regulation of human systems, on policy *objectives*; (b) strategic planning—focusing on general aims or needs formulated in terms of missions or functional envisaged outcomes, on strategic *goals*; and (c) operational or tactical planning—focusing on specific, attainable realizations, such as products, or operating *targets*.

Systems may then be classified as follows:

—*Rigidly controlled systems* pursue prescribed operational targets in prescribed ways for attaining them. (Examples: factories, bureaucracies.)

—*Deterministic systems* pursue prescribed operational targets but select between various ways and inputs for attaining them. (Examples: vertically organized product lines in industry, bringing into play various sets of material and nonmaterial resources, and in various sequences.)

—*Purposive systems* pursue prescribed strategic goals or multigoal patterns but select the corresponding operational targets (and the ways for attaining the latter). (Examples: Industry developing various and possibly innovative product lines; or diversifying in products and services under function-oriented headings such as "power generation, transmission, distribution, and utilization," or "food production, processing, and distribution.")

—*Heuristic systems* select their goals or multigoal patterns flexibly within the framework of a prescribed overall policy. (Examples: Industry developing new functional foci such as "environment," or "education"; universities setting up interdisciplinary programs.)

—*Purposeful systems* formulate and select policies in the light of the long-range outcomes (system states) of their own and their environment's potential dynamics. (Examples: New institutional roles for business, e.g., "planner for society," or for higher education, e.g., "education for self-renewal"; ecosystemic view of world dynamics and policies geared to stability rather than growth; etc.)

The above classification is of general validity and may be applied to human systems of all sizes and scopes. Examples may also readily be identified for nations or communities subjected to different modes of governance and planning, from rigid totalitarian systems to societies conscious of the necessities to regulate their own systems and open to cultural change (including changes in religious or ideological beliefs).

Figure 4 illustrates the evolutionary character of the proposed classification, indicating the stepwise realization of self-organization in terms of gradually bringing into play the processes at work in a fully developed human system. In this analogy, focusing on the self-organization of social systems, operational targets may be placed in the physical milieu, strategic goals or functions for society in

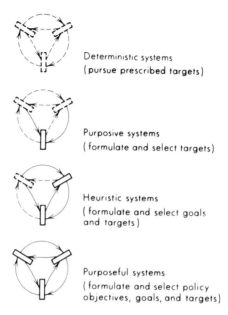

Deterministic systems
(pursue prescribed targets)

Purposive systems
(formulate and select targets)

Heuristic systems
(formulate and select goals
and targets)

Purposeful systems
(formulate and select policy
objectives, goals, and targets)

Figure 4. The external self-organizing characteristics of systems in terms of the process structure of human systems. (Dotted lines: rigidly fixed structures and processes; solid lines: alive structures and processes).

civilization, and policy objectives in culture. An operational target (e.g., a product) is formulated and selected by an instrumentality (e.g., an industrial corporation) in conformity with cultural rules pertaining, for example, to the utility or aesthetics of the possible targets and the outcomes of its realization. A purposive system formulates and selects a target, while keeping the strategic goal fixed. A heuristic system at the next higher step also keeps the goal flexible against the background of a fixed policy. Such a strategic goal (e.g., a function for society such as transportation or public health) is formulated and selected by the interplay of an instrumentality, bringing into play opportunities and potentials, with an institution (e.g., business or industry), introducing directions and constraints in conformity with the roles actually played by the institution. The purposeful system, finally, corresponding to a fully developed human system, also formulates and selects possible and desirable policy objectives, or overall principles regulating its dynamics. Higher-order systems may, of course, be composed of subsystems partly belonging to a lower level in respect to their self-organizing behavior; an industrial corporation, for example, highly flexible in its overall operations, may depend on a pattern of factories whose individual operations are kept fairly rigidly determined over some period of time. In Chapter 12 I shall return to this scheme when I outline a systems approach to planning, discussing it there in hierarchical terms.

Deterministic, purposive, and heuristic systems are all *teleological*, or finalistic, at different levels insofar as they pursue aims which, again at different levels, are inherent in the present dynamic situation. Only purposeful systems are *normative* in changing their behavior in cybernetic interaction with a changing environment in ways which are not necessarily prefigured by any aspect of the present.

There is a certain correlation between the internal and external self-organizing behavior of systems, although not an unambiguous one (therefore both views ought to be taken simultaneously in order to characterize a system). Figure 5 shows this correlation. It may be

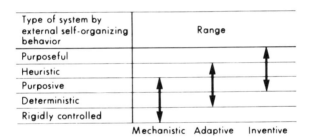

Figure 5. Correlation between internal and external self-organizing behavior of systems.

seen that human inventiveness alone can fulfill some of the tasks of operational planning in the realm of purposive systems—namely, the invention of new operational targets. Adaptive systems may only select from a given spectrum of targets but have considerable flexibility in doing so since internal organizational changes may accompany the selection of such targets. Mechanistic systems are restricted in the same task by the rigid organizational structure which has to fit all selected targets.

A deeper understanding of human systems may be gained if they

are viewed as incorporating various levels of natural systems. Laszlo
(1972 a) has attempted to describe natural systems—he divides
them into physical, biological, social, and cognitive systems—by the
same categories of systemic properties: systemic state property
(wholeness and order); system cybernetics I (self-adaptation, negative feedback); system cybernetics II (self-organization, positive
feedback); and holon property (intrasystemic and intersystemic
hierarchy). Summarizing what I have outlined in this chapter, I
believe that these categories can be ordered so as to bring out the
play between opposite emphases along basic systemic dimensions. In
Figure 6 I attempt to sketch these dimensions in relation to a hier-

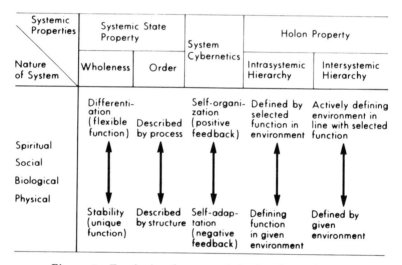

Figure 6. Emphasis of system behavior in various dimensions of systemic property.

archy of natural systems (physical, biological, social, and spiritual).
Such a representation provides a first glimpse at the complexity of
human system dynamics—because human systems are *not* just social
systems or spiritual systems but embrace the whole hierarchy. The
characteristics of physical, biological, social, and spiritual system
dynamics are all at work *simultaneously* in human systems—but
(and this becomes the key notion for understanding) in a *hierarchical order*. Thus, human systems face the task of *centering* in
the tension field mapped out by these dimensions. We cannot simply
rush from self-adaptation toward ever-accelerating self-organization
of widening scope, increasingly embracing also our environment.

As our social and spiritual systems become more dominant in the hierarchy, the life of human systems becomes more richly orchestrated across the entire range of natural systems, supported by "genetic continuity from below," and unfolded by "logical irreducibility from above," to apply some of Laszlo's (1972 a) core notions. In Chapter 14, I shall elaborate this theme further in developing the idea of successive "waves or organization" dominating the human world.

Most of the elaborate system models in use, and particularly in the economic area, are of a deterministic-mechanistic type, and few attempt to go beyond the purposive level. Most of the real social systems of Western society today function in a purposive or heuristic way (e.g., static or innovative "free enterprise" systems), usually modified or constrained not only by an overall policy but also by an instrumentarium of interventions which tends to become rigidified at lower levels. Purposeful systems are rare in the decisive institutions of society and hardly ever underlie planning efforts. Yet, in a period of cultural transition like ours, in which new values and new norms come into play and new roles for institutions (and possibly also new institutions) are gradually emerging, it is the highest step in self-organization which becomes of crucial importance for planning and actual change.

But human systems are not only devices for processing information and making decisions. They have a life far richer and deeper— a life which heightens, magnifies, focuses, and restrains the life of all the human members which express themselves through these systems. They have a consciousness which holds systems values and norms that may differ significantly from individual values and norms. They nourish aspirations and are capable of a creativity which differs from individual creativity. And they are capable of a feeling and insight transcending these faculties in the individual sphere. Certainly, we could not even speak of human systems if they were not more than the sum of individual expressions of their members. A social theory in a nonreductionist mood hardly exists today; it is probably not even possible on the basis of rational models of communication and creative action. Only if we understand human systems as instances of morphogenesis in an evolutionary process, can we reflect on purpose and direction in the course of human systems. Because the purpose cannot be the preservation of the individual randomly generated system—as the purpose of individual life clearly is not the preservation of the individual. But if human systems are alive in an evolutionary sense, they also play a role in design—perhaps a very important role since they constitute the highest evolved form which noogenesis has brought forth on our planet.

Our human systems have become too complex and too interrelated to be rebuilt in the happy artistic spirit which played such a prominent role in the design of temples, cathedrals, palaces, or whole cities in past times. The mutual enhancement of architecture and natural scenery through form and site in the design of Greek or Japanese temples, the classical harmony of the baroque abbeys in the Austrian Danube valley—which became the spiritual source for the Viennese school of classical music—or the joy, the melodies which seem to flow from the subtle blending of architectural variation in the design of Salzburg or Florence, are perfect physical expressions of such an artistic spirit in building human systems. But these were systems with human dimensions, systems whose parts could be directly related to human life, functionally a well as aesthetically, mythically as well as anthropologically. Most of the systems we are building today will be inhabited by people with technologically extended capabilities, functions, and desires. The modes of design appropriate for such systems will have to be more complex, too. They will have to combine reflective and nonreflective inputs, planning and intuition, intellect and *eros*. No unidimensional criterion will apply to the assessment of such design tasks—the abominable results of the application of crude economic criteria to large parts of urban America are there to prove this.

With new complex modes of design, where will the necessary creativity flow from? Will we have to set our hopes, as in some sectoralized modes of inquiry, scientific-conceptual as well as artistic, on individual creativity—or should we expect the emergence of some form of social creativity as it, to some extent, already characterizes more complex and interdisciplinary approaches to technology? Is individual creativity characteristic of nonreflective design, and social creativity of reflective modes? Will we thus have to develop new modes of design so as to ensure interplay between both? This would, indeed, become in itself an important design principle for human systems, implying that cultural, political, social, and generally communicative structures will have to be built into human systems accordingly. We have to allow for chances for individual leadership and imagination in participative systems; we have to design policies so as to ensure the inflow of creative imagination as well as a healthy evolution of these systems; and we have to design cultural bases which focus on creativity and communication rather than on egalitarian hedonistic ideals.

Question after question. We need a framework for ordering and reviewing them—perhaps answering some of them to our satisfaction. How does man become creative, how does he communicate with reality, and how does he conceive and set out to change it? Is this process enshrouded in mysterious fog, growing out of irrational depths—or can some of it be grasped and mastered more consciously?

Is the design process something which may itself become a matter of design?

In the next part of the book I shall attempt to sketch my model of the design process, of the human way of dealing with reality. It will underlie all further chapters which will discuss the various aspects of the design process and their significance for regulating the human world and, ultimately, the design process itself. But before I discuss the contours of the model, I have to explore the more profound issue of how we learn about the world in which we live, in what ways we relate to it, and what position we assume in it. In short, I have to discuss the subject/object relationship and the modes of inquiry which are based on it. In the next chapter I shall outline the three basic modes, or levels, of inquiry through which man relates to the world surrounding him.

PART II

Design: The Human Way of Relating to the World

Here, then, is a great mystery. For you who also love the little prince, and for me, nothing in the universe can be the same if somewhere, we do not know where, a sheep that we never saw has —yes or no?—eaten a rose . . .

Look up at the sky. Ask yourselves: Is it yes or no? Has the sheep eaten the flower? And you will see how everything changes . . .

And no grown-up will ever understand that this is a matter of so much importance!

Antoine de Saint Exupéry

IN the drama of existence we are, as Niels Bohr put it, ourselves both actors and spectators. Pretending to be only spectators, we have built a body of rational knowledge. But we live in a mythological world of relationships in which we are also actors. And in an evolutionary perspective, the duality of our role dissolves completely—we are agents of an evolution which acts within and through us. Design makes use of all three levels of perception or inquiry. It may be viewed as the ensemble of interactive processes between consciousness, a malleable "appreciated world" in our mind, and reality—and in three human spaces: physical, social, and spiritual (cultural) space. Such a ternary system of processes is capable of conceiving and implementing positive creative action—a notion irreducible to behavior which is the expression of a binary system characterizing most of the biological domain.

MODES AND LEVELS OF INQUIRY—REVIEWING THE SUBJECT/OBJECT RELATIONSHIP

The notion of design is often juxtaposed to the notion of inquiry, the former being considered the essence of artistic creation and the latter the essence of scientific research. At closer look, however, both appear as aspects of the same process which, in realizing its proper form, balances itself over time. I described this process of self-balancing in Chapter 2. The accents may be placed differently and hidden assumptions may suggest a difference in kind which leads us to believe in separate scientific and artistic modes of inquiry.

In our age such a conceptual schism is a surprise. We have come to recognize the processes which bring about creative advances in science, the new paradigms, as processes of human design, comparable to artistic creation rather than to logical induction or deduction which work so well *within* a valid paradigm. The gradual process of scientific theory-building and the verification/falsification sequence along a string of paradigms (Kuhn, 1962) may be understood as a creative self-balancing process no less than musical composition which Furtwängler (1954) described as "an 'improvisation perfecting itself' . . . in the balancing out (*Ausschwingen*) of its proper musical form, and yet in every moment, from the beginning to the end, remaining improvisation." Significantly, Furtwängler saw the norms of artistic design as being "inherent in the specific psychic process, by which a work of art is represented," and thus in the creative act, not in the created object—in the process, not the structure. This continuous self-balancing, this *Ausschwingen*, of a never-ending creative process is at the heart of truly great interpretations in music and the dramatic arts; creative and performing art evoke the same process and become identical in it.

This self-balancing of a creative process has been demonstrated most graphically in Clouzot's film on Picasso, in which the evolution of a drawing is followed through a long sequence of shots taken at various stages of the creative process. The subject is a Mediterranean beach scene and the first sketch shows a contemporary setting with a number of figures involved in joyful vacation activities. But then this familiar reality starts to change, acquiring a dramatic and almost demonic quality; the sea appears as a numinous source of life; the figures, now nude, seem to be engaged in sacred service; the spirit returns to the antique and even further back to archaic cultures. A female figure now occupies the center and grows, becomes ever more prominent—until life, all life on earth, seems to flow from the sea to her, and through her issues an expression of mythical dimension and power. But Picasso turns back to our time. Again the drawing becomes a playful beach scene—but something of that mythical world seems to vibrate in it; the simple joy of life has gained a deeper meaning. The evolution of the drawing is interrupted almost accidentally, the structure it has taken in the final picture appears arbitrary and without deeper meaning in itself; the meaning is in the process which, to the sensitive eye, goes on endlessly balancing itself out. "La peinture est plus forte que moi," Picasso once wrote; "elle me fait faire ce qu'elle veut." (The painting is stronger than I am; it makes me do just what it wants.) One may also think of the succession of drawings, etchings, and paintings in which Edvard Munch developed his great themes (e.g., the "sick child") over many decades. It is the process which speaks to us much more profoundly than any "definitive" end product.[1]

The basic process, the dynamic quality, in artistic expression touches us in all forms of art built to the measure and rhythm of human life, even in the world of stone and bricks in which architecture speaks to us. The subtle curvature of the Doric columns, the base lines and the temple walls of Athen's Parthenon draw us into a balancing process no less than the asymmetrical curvature of Viljo Revell's modern Toronto City Hall. Where development stops and structure is set before us in rigid blocks, as it is, for ex-

---

1. The current American preoccupation with "definitive" expressions and formulations—even "definitive" recordings of music!—is most revealing for a culture geared to static and technological modes of relating to the world. It also seems to imply that every expression, including the arts, is considered quantifiable.

ample, in certain kinds of modern music which reaches us mainly through block images of sound, we relate to it not in terms of life but in terms of a static cosmos.

The self-balancing of the artistic process becomes most explicit in the musical variation form. It is not the change in theme which adds most to the wealth and depth of musical development, but the qualitative changes introduced by exploring a theme from a wide variety of human angles of view. Beethoven's last quartets and piano sonatas, in which his human message is most intense and moving, have at their core movements written in variation form. Whereas the powerful juxtaposition of the strong "masculine" and the soft "feminine" themes in the sonata form introduces the tension present between opposites at the human level, the variation form lives out the tension inherent in man's existence between heaven and earth. Centering in these dimensions, and over time, is the secret of great art, as I shall discuss it in this book as the secret of all good design.

This same basic process, alternating between creative and appreciative phases and realizing its innate form, may be set in the framework of different approaches or modes of inquiry which are distinguished by the assumptions they make about the relationship between subject and object, observer and observed. In Figure 7 I

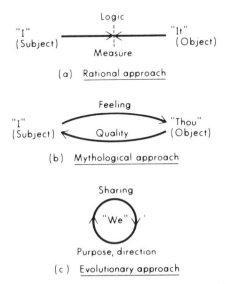

Logic

"I"   "It"
(Subject)   (Object)

Measure

(a)   Rational approach

Feeling

"I"   "Thou"
(Subject)   Quality   (Object)

(b)   Mythological approach

Sharing

"We"

Purpose, direction

(c)   Evolutionary approach

Figure 7.   Three basic modes of perception and inquiry.

have tried to sketch the different forms of relationships which underlie three basic approaches to inquiry:

—The *rational approach* assumes separation between the observer and the observed, and focuses on an impersonal "it" which is supposed to be assessed objectively and without involvement by an outside observer; the basic organizing principle here is *logic*, the results are expressed in *quantitative* or *structural* terms, and the dynamic aspects are perceived as *change*.

—The *mythological approach* establishes a feedback link between the observer and the observed, and focuses on the relationship between a personal "I" and a personal "Thou." Its basic organizing principle is *feeling*, the results are obtained in *qualitative* terms, and the dynamic aspects are perceived as *process*, or order of change.

—The *evolutionary approach* establishes union between the observer and the observed and focuses on the "we," on the identity of the forces acting in the observer and the observed world; the organizing principle is *"tuning-in"* by virtue of this identity, and the results are expressed in terms of *sharing* in a universal *order of process* (namely, *evolution*).

All three approaches are part of our world and are taken to elucidate different aspects of it. They constitute but partial aspects of a multifaceted subject/object relationship as it is evoked most suggestively in a thirteenth century Japanese Zen parable:

Two monks were arguing about a flag. One said: "The flag is moving."
The other said: "The wind is moving."
The sixth patriarch happened to be passing by. He told them: "Not the wind, not the flag; mind is moving."

Wind, flag, mind moves,
The same understanding.
When the mouth opens
All are wrong.

The *rational approach*, aiming at a detached view, corresponds to a phenomenological attitude which is not interested in purpose. It has divided the world—not only into disciplines, but also into a world amenable to objective science and an intractable metaphysical world, as Wigner (1967) points out: "The world is very complicated and it is clearly impossible for the human mind to understand it completely. Man has therefore devised an artifice which permits the complicated nature of the world to be blamed on something which is called accidental and thus permits him to abstract the domain in which simple laws can be found. The complications are called initial conditions; the domain of regularities, laws of nature."

Since the "initial conditions," which hide the complications, are assumed to have been set permanently once and forever in order to arrive at a fixed set of laws of nature, the rational approach is mostly geared to a static cosmos whose unchanging structure is unraveled by science—as a set of data to the positivists, as a structure to the structuralists. Science has not taught us much about the dynamics of processes, and practically nothing about the changes in structure which they bring about.

The distance between subject and object, between man and his environment, becomes possible through rational cognition. Man takes a position opposite to reality and imposes on the latter an intellectual order which exhibits the anthropomorphic features of his own rationality. He builds an abstract model of the world which follows the laws of his own rational consciousness and matches reality only from a specific point of view, impoverishes and narrows it down, as well as sacrifices many of its integral aspects. Man's physical, social, and spiritual space are divided into separate realms. The rational approach builds on the metaphysical assumption that the model is representative of the world, and that the unrepresented rest is of no great relevance. Scientific inquiry does not produce "objective truth"—it produces a human design which orders and formalizes certain aspects of reality in a communicable way. Kant emphasized the design aspect of science in his "Critique of Pure Reason" and his contemporary, Gianbattista Vico, ridiculed the naive belief in the universal validity and self-evidence of mathematical concepts which men understand simply because they have made them.

The use of the rational approach, set as an absolute, has had a devastating effect in the area of social dynamics. Either, as briefly discussed in Chapter 4, human systems are relegated to the domain of regularities, to behavioral science, or humans are taken to constitute or produce initial conditions and are not regarded as actors in a process. Contemporary social science is describing a Kafkaesque world in which the institutions and powers of society are anonymous and totally removed from the individual. But the rational approach also removes "the environment" and its regulation from the world of humans and human activity. It deals with man's world as being "artificial" and distinct from a "natural" world. The rational approach thus gives rise to a dualistic view, setting man against the world minus man.

The *mythological approach*, in creating a subjective relationship with the outside world, gains better access to an undivided, holistic reality which includes man. It explores the near-infinite spectrum of flavors the world holds in the psychic realm, the wealth of qualities which arise from our psychic response to the world with which we directly interact. This is the world of the "here and now" which is laid out in a finely woven web of qualities. Time acquires a direction here, but we relate to time through a string of momentary situations, through a succession of states of the web of qualities rather than through a direct sense of movement. In naming qualities we establish communication with the surrounding world, "get on speaking terms" with it, acquire the possibility of adapting to it and pleading with it—and, in a minor way, of influencing it in our favor.

The mythological approach corresponds to the *existential* view of life which concerns itself primarily with the conditions of man's captivity in a world which is "happening" to him. In the classic of modern existentialist philosophy, Sartre's "La Nausée," the basic existential experience is lived in the encounter with a stone. The experience of the material resistance in the stone generates a feeling of a specific quality, nausea, and thereby a personal relationship with the material world. Medieval alchemy tried to turn this passive relationship into an active one, where man would be capable of transforming the material world through psychic forms of interaction.

The process of seeing, according to recent research, is on the one hand an optical process yielding an image but on the other hand an energetic process involving the hypophysis and, through it, the entire person. In the light of these findings, the subject/object relationship becomes an integral feedback process in which both sides touch and shape each other. Goethe's *Farbenlehre* (theory of colors), put aside by science for a long time because it considers man's interaction with reality as such an integral physical/psychic process, is regaining our respect these days. Recent theories of cognition start from the basic postulate that the senses have a dual nature, a physical and a psychic one. According to Gosztonyi (1972) they form a field of tension created in the encounter between material processes and psychic receptivity.

Nobody has described with greater sensitivity and subtler psychology than Jean Giraudoux (1921) how the human mind, in tune with a reality full of psychic vibrations and life, is driven to create a mythological world of personal relations. In his novel "Suzanne et le Pacifique," he follows the inner adventures of an eighteen-year-old girl who is stranded in isolation on a South Sea island. She is a sort of anti-Robinson Crusoe. Unlike the latter who, in his scientific approach, carries his rigid European models with him wherever his fate leads him and finds nothing better to do than to build a dull replica of Europe in the lush, tropical life of his island, Suzanne is totally open and fluid—she virtually flows into the life of her island with all its fascination and lure, variety and color. She senses how life touches her there with an incredible wealth of facets, and she senses also that these forces try to caress and seduce her because, far from places where humans settle, "in this French girl, they see their only chance to ever become divine." A sigh in the forest begs to be named God of Silence, a rustling in the catleyas precisely at sunrise aspires to become God of the Red Ray or a flickering electric discharge in the sky just above the sunflowers God of the Green Ray, and a big tree, from whose inverted roots Suzanne always receives a rough blow on her shoulder possibly would like to become God of Caresses. But Suzanne does not yield; she gracefully plays with these forces but she remains suspicious of all this immodest begging and also of the more modest appeals addressed to her by aspiring half-gods. She continues to call the sea sea, and the wind wind and not Aeolus or Orpheus, and in addressing the island and the birds she says "vous" to put a distance between them and herself.

In our scientific and technological age, we are still living largely in a mythological everyday world whose order is built from subjective qualities and their interactions. The weather can be good or bad or fair, the sky friendly or threatening, a breeze strong or gentle; trees whisper and forests murmur, the sea rages or is calm; space and time are commodities which can be gained, saved, or wasted, and so forth. Our daily life is a series of interactions with objects which seem to treat us with friendliness or malevolence. It resembles the mime sequence in which the great Czech mime Ladislav Fialka encounters the objects of his personal surroundings, acted by two female assistants—the alarm clock which does not want to stop, the shower which yields only a trickle, the swinging door which kicks him in the back, the car which jumps, whereas at other times they all seem to outdo each other in impeccable, loving service to their master. On another level, institutions of society such as government or industry appear to some young people as personified evil, while they appear to others as the personification of justice or leadership.

Much of our political life is a play between such mythological figures. And our belief in a macroautomatism of ultimate self-regulation has mythological roots, too.

The mythological mode of inquiry is at the origins of both art and religion. In a hostile world, man set out to build an anthropomorphic world, a home, by placing himself in a network of subjective relations with the forces of nature and life, a world of personified actors which he was able to depict, speak to, bargain with, and venerate. It was not a predictable world of regularities, but at least one in which certain rules of conduct could be assumed. The world acquired a human face. But at its outset it was a world without a past or a future. Early cultures lived in the mythological "here and now," as the Navajo Indians of Arizona still do—and as children do in the first years of their life.

The *evolutionary approach* to inquiry, finally, not only considers psychic receptivity, but also psychic activity as an integral aspect of the evolutionary forces active in the world. "Das Äussere ist das in einen geheimen Zustand versetzte Innere" (The outer is but the inner transformed into a state of mystery)—Novalis's insight captures this sense of union between subject and object, between the observer and the observed. It is the sense of participation, together with the whole world, in the great order of process called evolution which makes it possible to learn about the universe by inquiring into our inner world. It is also the basis and justification of creativity issuing from this inner world; of our outgoing nature, of the active *yang* in us which acts in conformity with the passive *yin* listening to the pulse of evolution in us. Changing reality, then, transcends the mere feedback relationship and mutual adaptation of the mythological approach—in the evolutionary approach, man imprints himself on the world, shapes it according to his own image by virtue of his feeling himself an agent of evolution, of sharing in the essence of universal motion. Thus, the evolutionary approach corresponds to an *essential* attitude (as distinct from an existential one), interested in purpose and in the *primum movens*.

The transition from the mythological to the evolutionary level, from existential to essential attitude, is marked by the dissolution of the boundaries between self and the surrounding world. Witold Gombrowicz has caught this precise moment in which the static mythological world of personified objects merges with the observer, and both sides are drawn into the same overall rhythm of movement:

Henry (to a piece of furniture):
Are you looking at me? I am caught in a network of glances, in a
precinct of looks, and everything which I am looking at is looking
at me
    Even though I'm alone
    Alone
    Surrounded by this silence . . . I stick out my arm. This ordinary
    Normal
    Commonplace
    Gesture becomes charged with meaning because it's not intended
    For anyone in particular . . .
    I move my fingers in the silence, and my being
    Expands itself to become itself
    The seed of a seed. I, I, I! I alone!
    And yet if I, I, I alone am, why then
    (Let's try that for effect) am I not?
What does it matter (I ask) that I, I am in the very middle, the very
center of everything, if I, I can never be
    Myself?
    I alone.
    I alone.

(*The Marriage*, Act III)

The feeling of expansion, of becoming "the seed of a seed," is the
direct experience of evolution working within and through ourselves.
We become alienated from our subjective mythological world where
we had established ourselves in the center of a web of feedback
relationships. "What does it matter," indeed, once we recognize our
own being as a form of expression of an all-embracing evolution—
we can never be our isolated selves again.

Artistic design extends from the mythological level, with its
provision of a wealth of psychic responses cast into qualities, to the
evolutionary level where it acquires meaning as an instance of a
great overall order of process. An artist, says Ingmar Bergman, acts
like a radar, receiving the "things" of the world and reflecting them
back, mixed with memories, dreams, and images. Art is often mistaken as "mimesis" of reality, but great art of all times has changed
reality, given it a new structure. In our days, the best and most self-
evident example is probably modern architecture. But the same may
also be shown for the contemporary form of dramatic art which
is perhaps most alive, the film. An example appears in the key moments of Bernardo Bertolucci's film *The Conformist*—in particular
the scene in the dancing establishment in which the two women break

through all conventions and start dancing together a daringly erotic tango of such grace and beauty that everybody but the conformist feels joy, applauds, and joins in a gay polonaise. Such moment break up rigid reality and show it as open, changing and changeable, and full of chances for freedom. The tennis game without a ball at the end of Michelangelo Antonioni's film *Blow-up* is an expression *in extremis* of the creative interaction between a reality which is not ready-made and unambiguous on the one hand, and looking, which is an active shaping of reality and not passive sensation, on the other. Mime, the subtlest form of dramatic art, creates a world of qualities which becomes transparent to reveal a world of meaning, of essence. When Marcel Marceau, as "Bip" naively enjoying the beauty of an imagined butterfly in his hollow hand, suddenly realizes that the wish to keep beauty is its death, his whole body expresses this unexpected tragic insight and the world of senses and qualities acquires a new depth, becomes an image of the human condition. But it is perhaps in the ancient Japanese art of Noh, in the exquisite harmony of that unique blending of music and mime, dancing and chanting, movement and color, drama and ritual, that the pulsating rhythm of the inner human world and a numinous cosmos may be experienced most irresistibly.

Such a creative union between human and cosmic order is particularly striking in that kind of artistic activity which, on the outside, may be most easily mistaken for one-sided and passive depicting, instead of creating—photography; in fact, photography is often relegated to the realm of the crafts rather than to those spheres which we traditionally reserve for the arts. Yet the work of artists such as Wynn Bullock demonstrates beautifully and convincingly the dramatic quality of the artist's interaction with a reality which lies in processes, not structures. "What is material, tangible and permanent to most people is ephemeral to me," explains Bullock (1971). "An object is actually a visual or mental concept. It has no independent physical existence. Only events exist." This process between the artist and reality bases on a view of all things as being inextricably part of their opposites; only together they form a single relationship of meaning. In Bullock's work this principle becomes evident, for example, in his famous picture (exhibited at the Museum of Modern Art, New York) of the smooth white body of a girl, lying in the lush, finely woven textures of a redwood forest floor. The tension between the opposites makes an event out of what, on the outside, may appear as a static situation. Life-giving light filtering through decaying leaves has the same dramatic quality, transmitting a sense of finality and process. We find here the same theme of living out tensions which I identified with full life in Chapter 1.

The beautiful photograph by Angela Maria Longo, which forms the frontispiece of this book, may serve to demonstrate in a most suggestive way the subject/object relationship when different modes of inquiry are applied. What the camera depicted—and what a rational approach would endeavor to find out (with some difficulty, in this example)—is an arbitrary phase in some natural movement, as inconspicuous and ordinary as it occurs at the muddy bank of a river. Most people to whom I showed this picture felt obliged to try the rational approach, with vastly differing explanations ranging from "a surface of ice" to "a duck taking off." How irrelevant scientific "truth" appears here! A young and sensitive friend felt frightened, even hurt, by the picture—it struck her like an open wound, the pain of which she felt in herself. Taking the mythological approach, she established a very personal relationship with the quality of a form brought about by an arbitrary event. But the artist calls her photograph "Rising to the Sixth Chakra"—which, in the Hindu scheme of the seven chakras or levels of consciousness, stands for wisdom, represented by the "third eye." In an evolutionary approach to inquiry, human and natural movement flow together, physical and spiritual meaning become aspects of the same overall motion. In this mood, one may see a flame, a symbol of life, striving toward light—or, rather, the reflection of the divine light which we touch in wisdom. Light is also caught in the short-lived forms appearing at the edge of the flame, as it cuts into the smooth surface, realizing its own form; they may symbolize our passing models and myths which form at the interface of human mind and reality. And the small light in the upper right corner perhaps points to some of the basic truths to which wisdom will lead us.

Rational, mythological, and evolutionary modes of inquiry typify three aspects of design applicable to the building of an anthropomorphic world, a world in which man is at home. The subjective mythological level is suspended between two objective levels—the mechanics of events below, and the meaning and purpose in which human life is embedded, above. If we compare these levels with the types of systems I have outlined in the last chapter, as well as with the basic concepts of human and social dynamics, we find a striking correspondence which points to the terms in which understanding may be gained through these modes of inquiry:

Rational approach	⟷	Mechanistic system; behavior, causality
Mythological approach	⟷	Adaptive system; action, homeostasis
Evolutionary approach	⟷	Inventive system; regulation, homeorhesis[1]

I shall elaborate this correspondence further in Chapter 11.

This analogy underlines what is dawning on us in many different ways today—namely, that the rational approach is inadequate as an exclusive tool of planning for human systems and their regulation. The mythological approach with its subjective valuation underlies much of current politics and planning which, on its polished surface, is presented as the result of a rational approach. But it is only the evolutionary approach which allows the capacity for policy design to be fully brought into play for regulating the course of human systems as part of the stream of evolution. We are planning for a rational world, but we are mainly living and acting in a mythological world, and our actions impinge on the course an evolutionary world takes. Herein lies perhaps the most fundamental mismatch obstructing our intentions, ambitions, and goodwill, the basic calamity of our time. In rationalizing our world we tend to reduce action to behavior, quality to quantity, experience to data, values to measure, insight to empirical evidence. Without evolutionary inquiry we lack a sense of direction, without mythological inquiry a sense of systemic existence. Without both, we separate ourselves from the world in which we live.

We live and act in a specific human way mainly at the mythological level. The most adequate form of expressing relations at this level is by means of self-balancing feedback processes, as I described them in Chapter 2 and shall use them in developing my model of the human design process in the next chapter. However, the mythological world —and it is very important to perceive this clearly—is an existential world of the "here and now," of spatial relationships and, at best, local processes only. In such a world we are prone to *react*, rather than *act* freely. This is the reason underlying the ineffectiveness of planning lacking an evolutionary perspective.

Our situation is comparable to one in which we find ourselves in

1. *Homeorhesis* is the term proposed by Waddington (1970) for the preservation, not of a stationary state (as in homeostasis), but of a flow-process. Disturbances are counteracted so as to bring back the process, not to where it was when disturbed, but to where it would have progressed if left undisturbed. Many biological and social growth processes are homeorhetic.

a sailing boat. If we understand well the currents of wind and water —that is, the objective overall flow which sets the stage for our attempts to steer a particular course—we may use these very powers for our own subjective purposes. We may drift with the currents, react against their forces by repeatedly correcting our course, or even cross against them; there is a considerable margin of personal freedom. But it can be exploited only if we recognize the limitations set by the objective powers and if we *design* with great care and precision flexible ways of using them by periodically adjusting sails and rudder. In the same way, *free action* at the mythological level depends on the recognition and subtle regulation of powers acting at the evolutionary level, limiting as well as enabling human action. Freedom is a meaningful concept only in the presence of boundaries.

In a time of rapid change, such as ours, complex systems of inter-meshing feedback loops, with their inbuilt delays in the responses to action taken within the system, can be regulated only by *anticipatory action*. Planning is about such anticipatory action—but the naive expectation that it will become effective through rational concepts alone is self-defeating. Institutional change in the human world has occurred in very significant mutations all through mankind's psychosocial evolution—the shift from hunting and food gathering to agriculture, and further to task sharing, trade, organized settlement, etc. But these were changes in reaction to changes in the relations to the environment and the limitations coming into effect. This is a viable attitude in times of slow change; with the actual attainment of a widely felt "alarm level" it is then not yet too late to conceive and implement corrective action and massive restructuring. But it is a dangerous and self-defeating attitude in the face of a system dynamics which would let many factors overshoot the "alarm level" if action is not taken earlier, in an anticipatory way. Only an evolutionary perspective of planning, a deeply felt sense of movement and direction, will mobilize and focus the energy within us which will lead to those evolutionary mutations which planning in the context of change is ultimately supposed to structure and regulate.

Effectiveness of planning is linked to the proper use of energy. The corresponding concepts at the three levels are:

—At the rational level, we perceive energy as a *force* causing an object to move (causing a specific effect); force is strictly an *ad hoc* view of energy.

—At the mythological level, we give energy a *direction*; we may also say, we convert available free energy into *power*, or *action*. (In our egocentrism, we often speak incorrectly of the generation of power, or energy, or action—what a revealing hubris!)

—At the evolutionary level, we *regulate* energy flows and conversion processes between equivalents—energy, matter, complexity, information (negentropy), motivation, etc. Regulation, within a framework of human scope—in particular, regulation of the energy processes in the physical, social, and spiritual domains on our planet—is man's chief contribution to evolution.

These different views of energy are already implicit in the types of change processes mentioned in Chapter 4: force (or Aristotle's locomotion and quantitative change) at the rational level; role playing (qualitative change) at the mythological level; and syntony (generation/corruption) at the evolutionary level.

It is highly significant that the same three-level order of inquiry also emerges from modern physical science, as it explores concepts of time, process, and organization. Quantum mechanics (1926) shifted the focus of inquiry from the Newtonian force concept to concepts of order and disorder in physical systems and to the quantized interactions within them. The Heisenberg uncertainty principle points to the impossibility of scientific detachment and "objectivity" in such areas where the process of observation interferes significantly with the process to be observed, for example in the atomic and sub-atomic domains; it formalizes an aspect of mythological inquiry. General system theory, taking shape over the past three decades (Bertalanffy, 1968), attempts to generalize (mainly in verbal concepts) the principles of quantum mechanics for adaptive systems, extending the scope of systems inquiry from the physical to the biological and the social domain (with adaptive modes of self-organization in primary focus—see Chapter 4). The theory of relativity (1905) focuses on evolutionary inquiry, on the self-realization of events and processes in time and space; it has still not yet penetrated very deeply into our overall relations with the world, and in particular, our concepts in the realm of science.

Prigogine (1972, 1973 a, 1974 a) attempts a first synthesis in the physical domain. He distinguishes between the deterministic level, the thermodynamic level, and the level of dissipative structures which correspond precisely to the rational, mythological, and evolutionary levels of inquiry outlined in this chapter. Reductionism tries to explain all processes of life at the deterministic level (Monod, 1970) —and some microscopic aspects may indeed always be explained in a deterministic or probabilistic mode. But even in the physical sciences, a new notion of quality now emerges at the thermodynamic level—a quality which is irreducible to quantity.[1] And the macroscopic behavior of all dissipative (sufficiently nonequilibrium) structures in the universe, not just life, now turns out to be truly evolutionary, i.e., moving toward higher states of organization and not toward higher randomness, as a reductionist view would expect. In Chapter 3 it became clear that deterministic and finalistic explanations of evolution form part of a hierarchy of levels of inquiry.

The transition between the levels of inquiry, in physics as well as in social exploration, is characterized by symmetry-breaking processes (Prigogine, 1973 b). Going from the rational to the mythological, from the deterministic to the thermodynamic level, implies breaking the time symmetry and introducing the concept of irreversibility; time is given a preferred direction of flow. Passing from the mythological to the evolutionary, from the thermodynamic to the dissipative structures level, implies breaking the spatial symmetry between subject and object, observer and observed. This general breaking of spatial isotropy leads to a nonequilibrium world in which "order through fluctuation" can become an evolutionary principle. The evolution of a city, for example, raises demands in organization which are met by specific anisotropic communication networks, or patterns of activity (e.g., business, shopping, or residential centers). As observers, we assumed for ourselves anisotropy of time—the capability of making a distinction between the past and the future. At the evolutionary level we have to recognize now such a "historical dimension" in all dissipative structures, i.e., in the entire nonequilibrium universe including, but not restricted to, the processes of life. Subject and object both become partial aspects of the same overall process ensemble of evolution.

All three approaches may—and indeed ought to—be employed at the same time but with a clear understanding of their scope and limitations. The rational approach is due where within a certain framework limited in space and time, a mechanistic system may be

---

1. The same quantity of energy, for example, may appear in different situations in the form of qualitatively different particles. If quarks, in accordance with a recent view, are considered as basic modules of particles, their quantitative combination generates very different qualities of matter, or different expressions of the same energy.

assumed. Cost/benefit calculations and factor analysis still have their merits in the framework of well-defined operational tasks for which the realization process has been "frozen," as is the case for example in the operational phase of product development, or in building a hospital, or a part of a city once their functions have been determined. The mythological approach is due where the quality of life is to be assessed as a system of manifold relationships between humans and other human and nonhuman elements in a human system, such as an organization or a community. These relationships are highly subjective and often express themselves in personal feelings toward objects, structures, or patterns of services. During riots in American cities, the inhabitants of black ghettos burned their own houses—not with the purpose of blackmailing the authorities into providing new homes for them, but because they hated their old homes. Social indicators, or the rational approach, hardly are capable of furnishing valid indications for the interplay of qualities as they are experienced by individuals.

Where the rational approach elucidates quantity, the mythological approach brings quality into focus. Science has always endeavored to explain a quality, such as hot or bright or green, in quantitative terms—with some success in many areas. But Lord Rutherford's proud word that "quality is nothing but poor quantification" does not really touch the distinction between rational and mythological approach; it does not even hold in the "hard sciences" anymore. In a world of yet unrestrained growth, we are experiencing dramatic changes in quality more and more frequently as quantity increases within the same structure. If a hundred thousand cars in a city grow to one million, this affects not only the quality of transportation, but also generally the quality of life in that city. As the whole world approaches limits to quantitative growth, the qualities which the world holds for us, change drastically.

Many of these irreducible notions of quality seem to be innate in some form of movement, in a process and therefore not amenable at all to the static inquiry characteristic of the rational mode. The psychiatrist H. Burkhardt (1973) has recently pointed to the basic human need of alternating periods of opening up and of closure in man's relations to the world. The opening manifests itself in the establishment of sensual-erotic relations, of partnership with persons but also with the nonpersonal world; it is lived out at the mythological level. But it needs the structuring and stabilization through the *ratio*, the temporary closing in a rational mode of relating to the world. If this alternating rhythm, steered by fear and joy, becomes suppressed, the rational institutions surrounding man lead to a one-sided reinforcement of rationality which in turn (according to Burkhardt) leads to alienation and aggressivity. The naive expectation of overcoming aggressivity by rationality, by logical argument, is self-defeating—as are all one-sided concepts.

The evolutionary approach to inquiry focuses on movement, on kinesthetic experience in the widest sense—the experience of moving with an evolution of which all events and forms in the universe are manifestations. Thus, the evolutionary approach, to the extent that it can be formulated and formalized at all, is due where direction and momentum are to be assessed, or in other words, where the design of a policy comes into focus. With the observer taking part in the movement himself, being source and agent of this movement at the same time, the limitations due to the Heisenberg uncertainty principle—location and speed of a moving object cannot be accurately determined at the same time—do not hold for the evolutionary approach. In fact, location and speed become meaningless notions within a universal process which also unfolds within and through ourselves; what becomes manifest are deviations in momentum and direction.

The fragmented world of scientific disciplines is certainly of great value for the holistic mythological world of integral qualities. The dialectical approach and its generalization in the form of the systems approach (see Chapter 12) provides a method for elevating knowledge obtained by the rational approach for application in a mythological world. This is also the meaning of interdisciplinarity and of what, in Chapter 13, I shall call interexperiential inquiry. The transition from a systemic, more or less static mythological to a dynamic evolutionary world has not found a valid methodology yet; the term "trialectics" has been tentatively proposed (Arica Institute).

The dualistic separation between subjectivity and objectivity, deeply ingrained in Western thought, dissolves at the evolutionary level. The rational approach tries to be objective by virtue of not being involved—and by assuming a clockwork mechanism which ensures observable regularity. The mythological approach is subjective by getting us deeply involved. But the evolutionary approach is objective at a higher level, because we are not only involved in a process but are the process itself. In the rational world we deal with opposites in a dualistic "either-or" view. In the mythological world these opposites become inseparable pairs—thesis and antithesis, out of which arises a new synthesis. And in the evolutionary world, finally, the tension between these opposites energizes a pattern of self-centered processes.

The image of the life stream—or, more generally, the evolutionary stream—which I introduced in the beginning of this book, now may be viewed differently from each of the three levels of inquiry (see Figure 8): Taking the rational approach, we stand on one bank and

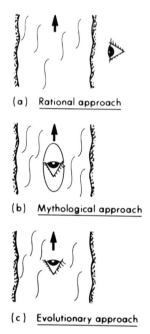

(a)   Rational approach

(b)   Mythological approach

(c)   Evolutionary approach

Figure 8. The three modes, or levels, of perception and inquiry illustrated by the image of a stream. At the rational level we are outside the stream, at the mythological level we try to steer our canoe in the stream, but at the evolutionary level we *are* the stream.

watch the stream flowing by; we do not get wet and proclaim our science to be "value-free." In the mythological approach, we are *in* the stream, drifting with it in a canoe which we may try to steer as best we can by relating to features which we perceive on both banks; we have no sense of direction and movement *per se*, but we respond to a series of match/mismatch signals emanating from the situations in which we find ourselves. But in the evolutionary approach we *are* the stream, source and flow, carrier and carried, the whole stream and yet only part of it—as a water molecule is the river and yet only part of it. Where the mythological approach attempts centering between opposites, the evolutionary approach seeks to center the process in itself.

The evolutionary approach embraces the mythological, which in turn embraces the rational approach. We find here a true hierarchical (stratified) relationship. Descending in this hierarchy increases the resolution of details and gives a better grasp of microscopic processes. Ascending gives increased meaning and order to the lower levels. In the evolutionary approach the tensions between opposites, which at times seem to tear the world apart, become the energizing force of a forward thrust. The zigzag between opposites which may frighten us in a mythological world, becomes a clear course in an evolutionary, finalistic perspective.

All three modes of inquiry contribute to our knowledge. Each of them has the same sort of trap built into it which marks the tedious way of becoming a "man of knowledge," as Don Juan, the old Yaqui sorcerer, describes (Castaneda, 1969): There are four enemies to overcome in learning. The first enemy is fear—fear of the overwhelming, seemingly amorphous mass of knowledge to learn. Fear is overcome by clarity which permits us to order knowledge. But clarity then becomes the second enemy which is encountered. Clarity is being content with a static representation of what one believes he knows, a model or an internally consistent body of knowledge as it is found in well-defined scientific disciplines. Clarity is overcome by power, by testing concepts in muddy real-life situations, by making them work and thereby losing their original clarity. Thereupon the third enemy is encountered—it is power, or the realization of great leverage which engenders the desire to act out power without wisdom. It is overcome by old age, by becoming detached and free of desires, less interested in the world than in the great principles which are at work in it, gaining control over oneself. We may also call it wisdom. So, finally, man encounters his fourth and last enemy on the way to knowledge—old age, physical tiredness, exhaustion from the long way. And if he still is strong enough to fight off this ultimate enemy, he may call himself a man of knowledge—for the short time that will be left to him.

Conventional science is giving in to the second enemy, to a treacherous clarity—a sweet defeat for the weak who have, so far, succeeded in making a publicly recognized and rewarded virtue out of stagnation. Technology is in the grip of the third enemy, power. For him who is driven further on his way to becoming a man of knowledge, scientists and technologists themselves have become the huge army commanded by these two enemies. Wisdom is not considered a virtue yet in science and technology.

In the following chapter, I shall explore the design process as a process of continuous learning through a multitude of interacting feedback relations linking ourselves and the world of our ideas to reality. To the extent that the learning process is kept open and alive in all of the loops represented in my simple model, it is capable of coping with the first three enemies encountered on the way to knowledge. The fear that befalls us when we relate to the infinite and overpowering wealth of reality is overcome by shaping a world of ideas—an appreciated world, as I shall call it—and making it an orderly image of reality. The resulting clarity, tempting us to settle on one specific image, is overcome by continuously testing the models and ideas for change—which make up our appreciated world—against reality and its demands. The sense of power, which arises from the recognition that through the appreciated world we can actually mold reality, is overcome by linking reality back to our own consciousness. It is the fourth enemy, old age, which cannot be countered within the design process itself—but in ensuring that there is an open and ever renewing sequence of individual consciousnesses and appreciated worlds, a sequence of generations, it gives the design process and man's striving for wisdom always renewed chances. In an evolutionary world, death is that aspect of life which pulls it forward, maintaining a continuous forward thrust.

(6)

THE BASIC DESIGN PROCESS—
MAN'S CYBERNETIC EXISTENCE

Design, in its full scope, involves inquiry at all three levels, the rational, the mythological, and the evolutionary. Viewed as a specific human capability, however, it is best defined at the mythological level. There, design appears as the core of purposeful and creative action, of the active building of relations between man and his world.

In this chapter, I develop my model of the human design process in terms of feedback relations, of a continuous dialogue between the three pairs which are formed of man's consciousness, the reality surrounding him, and the world of ideas, models, and plans he projects onto reality. The processes determining these relations may be viewed as evolutionary processes in themselves; I understand them as being of the self-realizing, self-balancing nature that I outlined in Chapter 2. In this perspective, the processes through which man relates to the world appear as a profound bond between design and evolution.

As outlined in Chapter 4, where I developed a typology of human systems, I view man's total space as being at least threefold: physical, social, and spiritual. The three aspects of man's space are often viewed in isolation, as if man related to reality through one of them at a time. Physical science and traditional epistemology have focused on the physical aspect; behavioral and social science as well as law on the social aspect; theology, and certain schools of philosophy, anthropology, and cosmology on the spiritual aspect. But it is the interaction between all three that sheds new light on movement and on the issues of regulation.

In Figure 9, I have sketched a general model of the processes involved in man's relating to his world, which I believe and hope to be supportive for the argument I intend to develop. It ought not to be taken for anything beyond an *ad hoc* model geared to a particular way of arguing.[1]

1. Only after I had finished the manuscript of this book, I learned from Sir John Eccles of Popper's (1972) three-world concept of human relations.

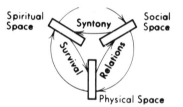

(a) View of the toroidal model "from above"

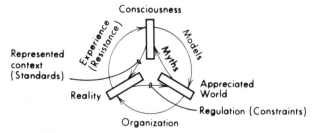

(b) General cross-section of the toroidal model
     (for each of the three spaces:
     physical, social, and spiritual)

Figure 9. Two views of the toroidal model of the basic human design process.

Viewed "from above" (Figure 9a), the toroidal model is composed of processes which interact in three slabs, representing man's physical, social, and spiritual space. Again, these three spaces do not represent any fixed structure, but may be compared to standing-wave patterns generated, and changed, by interference among various processes. The link between the physical and spiritual aspects focuses on *survival*, not only physical but also social and spiritual survival. It has to do with building a home in a cold, hostile world. Apart from ensuring the necessary physical supplies, man has a need for defining himself, his place in and his relations to the world. The establishment of iconological systems (as I called them in Chapter

Popper's World One contains "physical objects and states," World Two "states of consciousness," and World Three "knowledge in objective sense." My model differs from Popper's in two important ways though: My "World Three" which I call the "appreciated world," is built of subjective human models/myths, including the subjective models of science; and my model is toroidal, with each "world" interacting with both other "worlds"—whereas in Popper's model there is no interaction between World One and World Three. Eccles (1970), in his modification of Popper's model, establishes mutual interaction between all three worlds and views World Three, the world of culture, as made by man and making man.

4) such as the arts, mythologies, or language give expression to this need. The link between the physical and social spaces focuses on *relations*. They manifest themselves in becoming organized into human systems. The link between social and spiritual space, finally, may be labeled *syntony*. This notion, which I already mentioned when I discussed modes of communication in human systems, touches on the mystery of shared values and beliefs, of a shared sense of direction and impulse to advance, as it characterizes the behavior of societies within an effective cultural framework near the peak of a culture. Syntony, as it is moving us from within ourselves, is inquiry at the evolutionary level par excellence.[1]

The process I labeled "survival" has opened up man's physical space and been allowed to fill it. As I have briefly sketched in Chapter 3, relations—or social roles—have opened up social space which is about to fill up to the rim. Will syntony open up spiritual space now—and what will this mean? More people, more human free energy, or more ideas and information? In any case, the filling of a new space of human existence which is relatively empty so far, would certainly signify a major mutation in human systems and in the overall system of mankind. But so far we lack the imagination to envisage such a new regime of human systems in concrete terms.

It is always the *total* threefold space which is involved in the regulation of man's world, not just one of the partial spaces. This we forget only too easily. We try to regulate man's physical space, "the environment" as it is called today, as if we participated in it only in a physical capacity—as a land user, a dweller, a producer, or a polluter. Or we try to regulate man's social space by legislating and "building fences" around areas of common social concern. Even when we consider the interactions between the physical and the social space—for it is dawning on us that man's social activities endanger the viability of our physical living space—we usually keep the spiritual or cultural space invariant.

We consider the relations between man as a physical being and as a member of society within a given cultural framework. This, of course, simplifies things enormously since we cut out two of the three pairs of regulating links between the sectors of the total human space, namely those which involve the spiritual space. This may

1. Teilhard de Chardin employs the notion of syntony in the widest possible sense to describe the transcendence of humanity to superhumanity which will act like one personal being, an immense self, in which the billions of individuals would basically function with one mind by direct syntony and resonance of their consciousness.

appear permissible in times in which an unquestionable and unquestioned cultural framework governs the actions and attitudes, the anticipations and wishes of a vast majority of people within a cultural system. But if there is no such shared framework—or if the one officially recognized is no more viable—such a neglect implies leaving aside the most powerful regulating forces in the whole system. Without them, social reality becomes increasingly unmanageable, as we are indeed experiencing today. Daniel Bell speaks of the present "extreme disjunction between the culture and the social structure, the one being devoted to apocalyptic attitudes, the other to technocratic decisionmaking." The neglect of regulation in the spiritual sphere has become the foremost threat to our physical and social world as well.

Thus, my first thesis states that any issue concerning the design and regulation of the human world is inherently an issue of regulating his total space by balancing its physical, social, and spiritual aspects.

Viewing the model presented by Figure 9 "from above," as I have discussed it so far, generalizes the processes which I have already described in Chapter 4 as those forming human systems. Cutting the toroidal model at any place introduces new types of processes, as the cross section in Figure 9b shows in a general way. They link consciousness to reality and to an "appreciated world." All these notions constitute aspects of an evolving integral self.

With Roberts (1970) I view consciousness as "a dimension of action, an almost miraculous state, made possible by . . . a series of creative dilemmas." The first of these dilemmas arises from that aspect of self which forms the part of reality experiencing the evolutionary urge toward self-realization and the inability to do so completely—from it springs action which, in turn, forms identity. The second dilemma "between identity's constant attempts to maintain stability and action's inherent drive for change . . . results in the imbalance, the exquisite creative by-produce that is *consciousness of self.* For consciousness and existence do not result from delicate balances so much as they are made possible by lack of balance, so richly creative that there would be no reality were balance ever maintained." The third creative dilemma arises between consciousness' attempts to separate itself from action—which is inherently impossible but generates what I label here the appreciated world (and Roberts calls ego consciousness). Viewed in these terms, consciousness as well as the appreciated world—and, indeed, the whole ternary system of self—may be understood as evolving, mutating nonequilibrium systems.

Recent views (Fischer, 1971; see also Chapter 9) tend to regard consciousness as a continuous spectrum with a range of reflective states flanked by ranges of nonreflective (ecstatic and meditative) states. Even more suggestive perhaps is a model of consciousness as an infinite regression of levels only the first ones of which are explicitly reflected upon—beyond the physical level of rational science those levels at which we become conscious of morality, emotions, poetry and other holistic insights, and numinous qualities and experiences. Of yet higher levels, most of us become aware at best through inexplicable "flashes" reaching us and bringing our reflective levels to strange modes of resonance. Such a notion of consciousness goes far beyond what has been called discursive consciousness (explicit and analytical-reflective) and includes also intuition, spontaneity, direct insight, wisdom of the body, and—to the extent that it is active in the evolution of the whole system of self—all that is usually relegated to the domains of the subconscious and the unconscious. The evolution of consciousness would then be characterized by bringing higher levels into more active (more conscious) play in our lives, like an ever more fully resounding and more richly orchestrated chord.

An animal faces a reality which it cannot shape save in minor ways; principally, it adapts to the reality by means of a genetically preprogrammed feedback link of signal and instinctive response. The genetic program governing this form of communication changes only very slowly with biological evolution. But man has not only a much richer repertoire of interacting with reality directly—through broad qualitative "gestalt" experience instead of signals, through the formation of an explicit representation of his existence and its circumstances, and through the possibility of massive and effective transformation of reality by way of human action. His system of relating to reality includes the possibility of bringing into play alternative options, of anticipating and acting in line with his anticipations, which include inputs from his own will, of reflecting on reality and his views of it, of organizing—in short, the possibility of *planning* and *acting with a purpose*. In broader terms, we may say that man has a certain capability of *designing* his world which transcends the possibilities of other creatures on earth.

This design capability may be viewed as working essentially through something like a hydraulic control system mediating between man and crude reality. Its "pressurizer," in this image, would be what Vickers (1968 and 1970) calls an *appreciated world*.[1] This appreciated world came into being with the development of man's capability for self-reflection, a faculty encompassing much more than just thinking. It holds the world—the physical, social, and spiritual aspects of man's world—as we view it not just through the understanding which our mind composes of it but through all forms of experience. It embraces our appreciation of what this world can do to and for us, and what we can do to and for it. It reflects our own place and movement in the world as well as our responsibility toward it, the demands which we make on it, and the personal concept we have formed of it. Most importantly, it holds the differences between the world as we want it to be and the world as we actually perceive it. Thus, the appreciated world becomes the motor for change induced by human action.

In its relation with the appreciated world, our consciousness forms one integral system with it. The appreciated world is part of ourselves as persons. In the appreciated world, technological man is not the naked ape which picks up tools and lays them down after use—he is man with his senses, capabilities, and expectations extended by technology, man with the mechanical force of huge machines, man traveling at the speed of sound and beyond, man reaching with his voice around the world and into space, man processing huge amounts of information, a monster acting with incredible leverage on physical and social reality—and, unconsciously, on spiritual reality, too. He is also man with his desires and anticipations, an active element in a dynamic world, and he is man who is becoming conscious of the systems of human life which he populates—in short, he is man the *cybernetic actor*.

The appreciative system is man's unique device[2] for relating to a

1. Quite generally, the model I present in this chapter may be understood to a considerable extent as an extension and elaboration of basic notions which I have learned from Sir Geoffrey Vickers and which have become so fruitful for my own thinking. Of course, I assume full responsibility for their adaptation to my own arguments.

2. When I say "unique," I mean among living creatures on earth. Actually, the extension of a binary subject/object relationship to a ternary relationship including a malleable building and testing grounds for ideas and complex systems of ideas, may generally be a principle underlying creativity at all levels

reality in whose shaping he is actively and creatively participating. It marks the appearance of culture. Not only is it, as Vickers (1970) states, "carved out by our interests, structured by our expectations, and evaluated by our standards of judgment"—corresponding to the arrows directed toward the appreciated world in Figure 9, and called "regulation," "models," and "represented context"—it is, even more importantly, also the generator of myths on the one hand, and of organizing activity on the other. Both myth formation and organization tend to stabilize the relations between ourselves and the world in a static mode, creating platforms for elaborate and powerful action which then turn into strongholds of resistance to change.

Myths supply us with a frame of reference, with some basic overall principles which provide meaning, something we would like to maintain while we are in the process of molding a specific appreciated world. The image of a hydraulic system stands for an ideal which will never be fully realized because we find it so difficult to manage an appreciative system with considerable delays built into it. It needs time to translate an appreciated world into the organization of a real world, and it also needs time to reflect on consequences and constraints, on issues of regulation. Man's capability of managing intermeshing feedback loops with long delay times is limited, whether the system in question belongs to the outside or to his inner world. Therefore, the ideally fluid process tends to become a batch-type

---

in a very profound way. We may speak of *autocatalysis of consciousness*. This seems already realized in a passage from Lao-Tzu's "Tao Teh Ching":

> Tao gave birth to One,
> One gave birth to Two,
> Two gave birth to Three,
> Three gave birth to all the myriad things.

A binary system works primarily through negative feedback (adaptation) or behavior; but a ternary system is capable of introducing positive feedback, or active design, in highly differentiated ways. In Hermetic and Whiteheadian philosophy, the All (the "extensive continuum") creates the evolving universes in the same way by first conceiving, building, and perfecting a "plan" which subsequently is implemented in the evolutionary process in reality. The same image also recurs in many instances of direct insight and intuition, whether they become then formulated in scientific terms or not. An interesting recent example is the description of creation in the "Seth material," a body of revelatory knowledge received through a psychic medium (J. Roberts, 1970): In giving actuality to all the probabilities he had developed and elaborated in his dreams, God lost that portion of himself which had created them. His "appreciated world" became reality. By this process, consciousness multiplied itself and spread creativity in reality.

operation—building a particular appreciated world and testing it against reality, building on the basis of another model another appreciated world, and so forth, a "stop-go" process which only too often degenerates either into stifling rigidifications or into an increasingly unbalanced stumbling from one static platform to another. The usual way to operate the "stop" in this process is to set an absolute, unchallengeable myth. The opposite does not work as easily, as is demonstrated in particular by the resistance to institutional change within a cultural system.

That myths are generated by man's appreciative system, not through his direct links with reality, is of vast consequence. It means, first, that *myths are necessary*; together with models, they form the basic cybernetic loop between man and his appreciated world, through which he reaches a changeable reality. There is nothing bad about myths except their tendency to become absolute and lasting. A rigid myth stifles the creative process of forming other appreciated worlds, of communicating with reality in a vivid way. Secondly, it means that *myths are a matter of design, too*. Not only can they be designed in conformity with any appreciated world—or as a formative principle for any appreciated world—but they can also be designed so as to "tune in" with reality, to become powerful regulators in designing man's physical, social, and spiritual world in a viable way.

I call the operation, in man's threefold space, of the three pairs of feedback relations between man, his appreciated world, and reality, the *human design process*. My second thesis concerning the design and regulation of the human world is that these are effected by means of the full design process; in particular, design incorporates the building of appreciated worlds through model and myth formation in a feedback relationship and the linking of alternative appreciated worlds to reality by means of organization and regulation.

The three pairs of feedback loops constitute but a crude model, a first approximation. It would be misleading to imagine that they are "operated" in this simple way, running to and fro all the way between a pair of pivots. Rather, the processes involved might be viewed as being of the general self-balancing type which I have described in Chapter 2, prancing back and forth in little steps, or running through tiny feedback loops within feedback loops within feedback loops.

The planning process, as I view it, embraces forecasting as its innermost core and is itself embedded in the decision-making process and generally in the process of rational creative action (Jantsch, 1972). It now becomes clear that the planning process forms an integral aspect of the design process, working in the same way and in the same pairs of feedback cycles. Planning may perhaps be viewed as the rational aspect of design—the latter being the broader notion and including also aspects that, so far, we would relegate to realms beyond rationality. But at this stage, I prefer to let the notions of planning and design merge completely, at least at the mythological level. It would be futile to restrict the preparation of purposeful action to planning in a merely rational sense, and let the aspects of quality and personal experience elude it. In the next chapter, and again in the Epilogue, I shall reflect again on the notions of planning and design but viewed from the evolutionary level of inquiry. There, planning will appear as complementary to love, and both notions together as integral aspects of design.

In terms of the planning process, the feedback link between man and reality would refer to the assessment of a situation, together with an inducement to devise some action for change—a social program, or the development of a new technology, a product, or a service. The feedback link between man and his appreciated world would correspond to forecasting at the three planning levels, as I have described it elsewhere (Jantsch, 1972)—predicting specific consequences at the operational level, inventing and elucidating options at the strategic level, and restructuring norms at the normative level. The feedback link between the appreciated world and reality finally would correspond to the anticipatory evaluation of potential action, to the projection of feasible and desirable action onto reality, the harmonization of the appreciated with the real world, and the matching of potentials with constraints, those in material and nonmaterial resources as well as those inherent in the dynamics of real systems.

Thus, the planning/design process "embraces" reality from two sides, thereby giving man a unique flexibility in dealing with it, in adapting as well as imposing himself. That the planning process also touches simultaneously on all three sectors of man's total space—physical, social and spiritual—is only now becoming gradually recognized and has not yet made much headway in actual application. One of the major obstacles to such an expansion of the planning concept is the inclusion of man's spiritual space—which, of course, leads planning out of the purely rational domain. But this is a fruitful dilemma because it can only act as a spur to promote the renewal and expansion of concepts which are becoming less and less viable. A closer look at the particular meaning of the feedback relations within each of the three sectors, as illustrated in Figure 10, brings

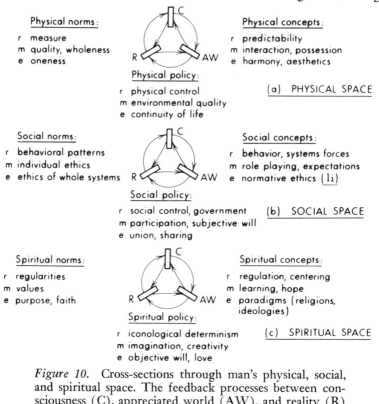

Physical norms:
r measure
m quality, wholeness
e oneness

Physical concepts:
r predictability
m interaction, possession
e harmony, aesthetics

Physical policy:
r physical control
m environmental quality
e continuity of life

(a) PHYSICAL SPACE

Social norms:
r behavioral patterns
m individual ethics
e ethics of whole systems

Social concepts:
r behavior, systems forces
m role playing, expectations
e normative ethics (li)

Social policy:
r social control, government
m participation, subjective will
e union, sharing

(b) SOCIAL SPACE

Spiritual norms:
r regularities
m values
e purpose, faith

Spiritual concepts:
r regulation, centering
m learning, hope
e paradigms (religions, ideologies)

Spiritual policy:
r iconological determinism
m imagination, creativity
e objective will, love

(c) SPIRITUAL SPACE

Figure 10. Cross-sections through man's physical, social, and spiritual space. The feedback processes between consciousness (C), appreciated world (AW), and reality (R) give rise to norms (standards for regulation), design concepts, and policies. The focus of these key notions of the design process depends on the level of perception/inquiry : r = rational level (change), m = mythological level (process), e = evolutionary level (order of process).

interesting facets of the human design process to light. By absorbing and processing experience, and projecting it back onto reality in the form of representations of the context of human existence—Who am I and what is my place in the world?—we start building an alive feedback relationship with the world. We experience, not just the world, but ourselves in this world. Out of this interaction with reality evolve the *norms* which we recognize for our existence—

they are the fruit of man's most direct reality-learning process, of experience and self-experience. These norms appear in different aspects corresponding to the rational, mythological, and evolutionary levels of perception and inquiry. In Figure 10, I spell out these aspects.

It is of great significance that, in this model, norms, or standards for regulation, develop not by consciously molding the appreciated world, but evolve in the direct interaction between consciousness and reality. This is consistent with Vickers's graphical notion of a "tacit norm" (Vickers, 1969), a screen of schemata through which we filter our experience and which is itself a product of the process it mediates. Alexander's notion that design is essentially the elimination of "misfit" as the designer proceeds by measuring tentative design ideas against tacit criteria, mounts the same horse from the opposite end (Alexander, 1967). "Tacit norms," gained in direct interaction with reality, guide the design of an appreciated world in another feedback loop, which also involves our consciousness.

I have already emphasized the paramount importance of the feedback loop formed by *models and myths*, that link man to his appreciated world. A myth is any model we believe in and let govern our attitudes, design, and activities. In this sense conventional science has become a rational myth in our days—the myth of the one absolute truth which the indefatigable ambition of man continues to unravel. Again, qualitatively vastly differing aspects are expressed by myths pertaining to social space (e.g., the "class struggle") and myths pertaining to spiritual space (e.g., "humanity is one spirit"). Cosmological myths, in the form of religions—and also ideologies—govern cultures and determine to a large extent the social myths that are deemed "admissible."

Every religion and ideology includes a prescriptive ethics and very conspicuous examples of historical myths may be studied in our days in their dependence on totalitarian ideologies. Much planning in the social realm presupposes that any ethics can be easily altered if the "good" or "bad" of certain courses of action, and according to shared criteria, can be unmistakably demonstrated. The "limits to growth" were not difficult to grasp—and yet, growth has not become limited to any extent in any area, neither in industrial activity, nor in agriculture, nor in population, nor in income and wastage. The cosmological myth of the self-regulating macroautomatism of the human world, the liberal religion, eagerly defended by narrow-minded economists, overruled any concern emerging at the social level. Here, we already see at work a certain hierarchical principle, to which I shall return later.

We "test" our appreciated world against reality by a feedback cycle alternating organization and regulation, from which emerges *policy*, the basic themes underlying our design efforts. Physical organization bases on technology, perhaps the most conspicuous fruit to grow out of man's appreciated world, and his means to emancipate himself from reality sufficiently to make the whole design process meaningful and realistic. Social organization, or social technology, as one also hears it called today, manifests itself above all in the central concept of human will, or role playing, to which I shall return in later chapters. It forms the core of a social theory based on action instead of behavior. Spiritual organization finally has much to do with "objective will," or love, with the spiritual force by which man is able to touch the reality that is formed by the consciousness of humanity and the spirit in which humanity participates. Therefore, instead of "spiritual reality," we may also use Teilhard de Chardin's term "noosphere," which denotes the consciousness of all consciousness, centered beyond humanity in an evolutionary *telos*.

Regulation is the fruit of careful reflection. I have already pointed out that the development of a capability for individual self-reflection in man was the prerequisite for his forming an appreciative system, and thereby for the design process as such. The appreciated world evolves from a dialectical process between organization/regulation and the utilization of consciousness through models and myths. In the last phase of his long evolution, the forty or fifty thousand years of psychosocial evolution, man has also developed the capability for collective reflection which indeed was necessary to arrive at the possibility of higher social organization or civilization. In spiritual space the corresponding notion would be evolutionary regulation, which would amount to the possibility of consciously designing cultures. Mankind is not there yet; evolutionary reflection is at work in but a few individuals, in spiritual leaders, saints, artists, great scientists. This last and crucial link is still missing like a tooth missing from a cogwheel and thus removes the overall design process from man's full awareness, blocking it and bringing it into gear with unforeseeable mutations that frighten us with doubts and throw us back to hope or despair. Evolutionary reflection may be regarded as a step in the evolution that mankind has not yet made but is preparing for. However, in this book I am trying to improve the understanding of this basic potential.

It is not accidental, that the clockwork arrows in spiritual space include the old triad faith-hope-love. In one of his mystical poems, Charles Péguy has compared faith and love to the two big sisters who take hope, the little one, in their middle; yet it is hope (on the more human mythological level) that leads them along their precarious path. This is a beautiful image for the regulation of spiritual space. Faith and love stream to us from our "tuning in" to reality—they nourish and protect a fragile hope that is born in man as he builds his appreciated world. With hope, man's path, his tao, becomes an evolutionary, self-balancing process.

The overall design process regulates man's consciousness, his appreciated world and reality. But how is the design process regulated, or regulating, itself? In the next chapter I shall try to throw some light on this crucial question.

(7)

A SENSE OF MOVEMENT AND DIRECTION–CENTERING THE DESIGN PROCESS

In the last chapter I sketched the basic human design process in a way which left it open whether it might also be envisaged as unfolding between entities or aspects of the world, some of which may be assumed as fixed. This would permit the design process to follow a logical procedure, starting, say, from a fixed reality or from man's consciousness. Both reality and human consciousness are "hard" notions in Western thought, and only the appreciated world would then represent a malleable element through which we might try to change reality in minor ways, to "reform" it.

Western man has a strong bias in favor of identifying an Archimedean point from which to gain leverage over the world—in fact, he has a strong bias for the idea of leverage as such. He finds his security in firm structures, physical as well as social and cultural ones, not in balance of movement. The only way he knows how to deal with the world is sequential, starting from firm grounds and proceeding step by step in solving what he calls "problems." He abhors the idea of being afloat, of abandoning himself and of feeling carried; he trusts only petrifacts and is suspicious of things alive. He even likes to build his appreciated world as firmly as he can.

But man's basic relation to reality does not resemble a firm grasp but rather a circling or balancing out in space and time. When a child in a baby carriage is given a toy or any object, its first impulse is to throw it out, only to want it back the next instant. It is not the object which is real to him but the processes of gaining and losing. A child lives fully in a mythological world of processes, of experiencing quality, not yet within boundaries and structures; as it lives these processes, it lets them define the boundaries and structures of his world. It is only later, through schooling, and generally through all kinds of socialization, that structures are imposed from outside on his world.

Man has a biological need for processes, for the changes in light that go with the rhythm of the day and the resulting changes in his ways of relating to the world, for the continuously renewing processes of "building" his world. Certainly he is able to relate only to a reality which is structured in some way, which manifests itself in spatial and temporal rhythms. He needs structure to form himself —but he does not need any firm and unchanging structure; on the contrary. He needs a feedback framework of challenge and response to act out his full capabilities.

I drew Figure 9a (see p. 102) in such a way that man's total space forms a self-balancing system in which no part remains fixed. I also briefly discussed the difficulty of implementing such a regulation process; what usually happens is the "freezing" of spiritual space, of a dominant culture and its norms. Such a "freezing" stifles the design process. Self-balancing here means that every space balances itself between, and with the help of, the two other spaces—social space balances itself between physical and spiritual space, spiritual space between physical and social, etc.

But the full operation of all three feedback loops in Figure 9a is not sufficient. The same type of free self-balancing characterizes the full-scale design process also in the system formed by reality, consciousness, and the appreciated world. This implies that reality as well, even physical reality, cannot be assumed as fixed. It is not only subject to changes due to biological evolution, or to man's physical interference, but also to the evolutionary movement arising from physical reality and led by man in interaction with this reality. Reality becomes a dynamic notion.

In Figure 9b, in the cross-section of the design process model, we would then have two pairs of feedback links defining and regulating in each case one of the constituents of the system:

—Consciousness is defined and regulated by the feedback links to reality (represented context/experience) and to the appreciated world (models/myths);

—The appreciated world is defined and regulated by the feedback links to consciousness (myths/models) and to reality (organization/regulation); and

—Reality is defined and regulated by the feedback links to man's appreciated world (regulation/organization) and to consciousness (experience/represented context).

We may say even more aptly that consciousness, appreciated world, and reality are in a continuous process of becoming *centered* in the design process. No one-sided position can be maintained in this self-balancing of a system in which the processes are kept alive and support and check each other. The overlapping arrangement of the basic relations (processes) in the system, with plenty of possibility for redundancy, sees to that. Like all life processes, the design process has a great deal of redundancy built into itself. Indeed, such centering of states of consciousness, of an appreciated world, and of reality seems to follow the same principle of self-realization which underlies the evolution of living species, as well as artistic and other creative processes. The aim is always to secure a viable form, not by anticipation and straightforward realization of a specific structure, but by balancing it out—letting it define itself.

But centering, in the design process, is not a static or steady-state notion. The ensemble of feedback processes which constitute the complex design process takes part in an evolutionary, asymmetrical movement with a distinct direction. It is in this overall movement over time in which centering is sought—centering in the relations to a changing environment as well as in the relations between changing forms of consciousness, appreciated world, and reality. The centering of the design process transcends internal or external homeostasis, or adaptation to developments which are independent of the design process itself; it encompasses the active, deliberately "unsettling" nature of the design process seen in an evolutionary perspective.

If we take a closer look at the cross-section of the design process, we find a most interesting *asymmetry* in Figure 11. If the feedback

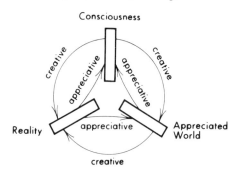

Figure 11. The asymmetrical scheme of centering in the design process.

processes linking reality, consciousness and the appreciated world are properly labeled as creative or appreciative (inductive or deductive, yang or yin), we find the following pattern: they are symmetrical with respect to reality and consciousness—the creative processes go out from consciousness in both directions, and impinge on reality from both directions; and the appreciative processes go toward consciousness from both directions, and emanate from reality in both directions. They are asymmetrical with respect to the appreciated world—the creative processes go from consciousness through the appreciated world on to reality, and the appreciative processes in the opposite direction, again through the appreciated world. This asymmetry is an expression of the restless, evolutionary, and "crisis-provoking" role of the appreciated world that does not swing in harmony with a static system and around a stable state but becomes a driving force for unsettling the human world in all its aspects. What becomes manifest here is the symmetry-breaking characteristic of the transition from the mythological to the evolutionary level of inquiry to which I pointed in Chapter 5. It is the symmetry between subject and object which is broken here.

Figure 12 points to other interesting asymmetries in the processes acting between reality, consciousness, and the appreciated world. In physical space, reality acts strongly upon consciousness whereas the latter has only a weak direct impact upon reality (which the rational approach negates altogether). We see physical reality through our eyes, but we can transform it only marginally and mostly in immeasurable ways in this process. Physical conventional communication theory, therefore, deals only with one-way processes from source through channel to receiver, not with the feedback influences which the receiver may enact on the source. But in the physical domain consciousness acts strongly on the appreciated world and through it on reality, with technology as a powerful mediator. Reality is almost helpless against the onslaught of imagination and potential innovation—except for the kind of delayed backlash which seems to be preparing now in response to rampant technology. This asymmetry implies that in physical space we acquire leverage through imagination, ideas, and planning; we find it easier to act upon, than to listen to, reality. Thus, the operation of the appreciated world and its infinite metamorphoses gives man the power to transform physical reality.

In spiritual space, the sense of strong interactions is the reverse. Spiritual reality, our "Weltanschauung," dominates the appreciated

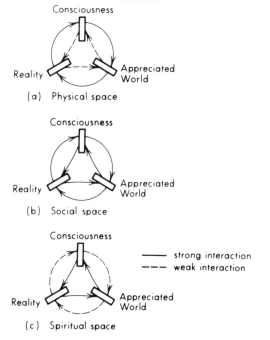

Figure 12. Interactions within man's different spaces. They are asymmetrical in physical and spiritual space but generally symmetrical in social space.

world, restricts its independence and life, and reaches further to impose a certain structure upon human consciousness. But in the spiritual domain, consciousness also has the potential of acting relatively strongly upon reality in a direct way—a potential which is tapped only partially and occasionally in a wide spectrum of psychic phenomena which occur largely outside the restricted realm of science. There is no communication theory yet of spiritual space, although there is undeniably communication such as telepathy, but also the continuous evolution of language, which plays a significant role in our daily life. I have already mentioned the concept of syntony, which seems to be due to direct communication of individual human consciousnesses through some resonance process; even plants, as recent research results demonstrate, are capable of "tuning in" to human consciousness. An evolutionary view would assign such communication its proper place in the overall picture.

Interaction in social space is more or less symmetrical—all interaction is of a relatively strong character. This makes social space the most malleable as well as the least regular one. As individuals, we find generally sufficient freedom to move, but we cannot escape being moved as well. Every personal interaction is clearly two-way; no source can communicate without being, at the same time, also receiver. There is no mere duplication of form but almost always interaction, which leads to instability and resonance in which source and receiver change on practically equal terms.

The balancing out of these processes works partly through match/mismatch signals according to standards which may be set almost onesidedly in the asymmetrical patterns of interaction in physical and spiritual space. But in social space, these standards are themselves subject to a subtle self-balancing which always involves both ends of a process. This is perhaps the reason for the high degree of freedom that the evolutionary process finds in the social domain.

Vickers (1970) has suggested the image of a football (soccer) game in which self- and mutual expectations generate subtly regulated standards; if player A gets the ball, player B starts into a free space expecting A to direct his pass there. Thus, evolving situations are judged by the same standards. Similarly, the formation of higher forms of social organization finds open a wide range of options, including such options which lead to thwarting the whole system of man's total space. This, one might speculate, may point to the necessity of active regulation in the social domain even more than in the others. We may perhaps put it this way: where action dominates in physical space, and listening to evolution in spiritual space, regulation—the subtle blending of both—is the major challenge in the social realm. Which, after all, is also what our intuition keeps telling us.

In an evolutionary image of design, it is not possible to start from a fixed reality, next build consciousness, then an appreciated world, and finally explore possibilities for "reforming" reality. This would be the conventional view of planning which starts from firm structures and assumes not only a fixed physical reality, but also a brittle social reality and on top of it a spiritual reality (a culture) hard as rock. It has led to such ineffective and short-range approaches as incremental planning, political bargaining and, in the words of the Bellagio Declaration on Planning (Jantsch, ed., 1969), generally to approaches that make "policies which are inherently bad—more efficiently bad." This is not design—and I would also not call it planning, except in an operational sense.

A description of the design process in terms of a system of feedback processes, as I have used it in the preceding chapter, emphasizes the mythological level of inquiry. In such a description, balance is only definable in static, or steady-state, terms. Every communication, which is equivalent to a transfer of energy, would have to be perfectly symmetrical. I have already pointed above to asymmetries. Also, in order to act, we need criteria to select from various options which are open in terms of models/myths or plans for organization and regulation.

Thus, the model of the design process has to be placed into an overall movement, through which design and action resulting from it are given a specific direction. Mythological inquiry has to be complemented by evolutionary inquiry, by a sense of movement and direction which acts within ourselves and which we impart to human design. This sense of direction (with which I shall deal in more detail in Chapter 9), provides the criteria by which to choose from a variety of design options. This "tuning in" to the overall movement, to *tao*, transcends the processes of communication between consciousness, the appreciated world, and reality. Its spirit is grasped beautifully in an old Japanese Zen writing:

> Joshu asked Nansen: "What is the path?"
> Nansen said: "Everyday life is the path."
> Joshu asked: "Can it be studied?"
> Nansen said: "If you try to study, you will be far away from it."
> Joshu asked: "If I do not study, how can I know it is the path?"
> Nansen said: "The path does not belong to the perception world, neither does it belong to the nonperception world. Cognition is a delusion and noncognition is senseless. If you want to reach the true path beyond doubt, place yourself in the same freedom as sky. You name it neither good nor not-good."
> At these words Joshu was enlightened.

If reality is conceived as dynamic, the design process may be compared to an attempt to climb a steep sand dune. Alone, this often turns out to be futile; with nothing firm to step on, one foot loses the same elevation which the other one is about to gain. Two people linking hands can often manage—with the same loose sand to step on which makes a single person despair. It is a process which does not involve any step-by-step structure—and any rest would ruin the whole effort instantly. It is a process of mutual balancing in dynamic movement, like running up on the surface of the sand

and using the other person's dynamic momentum to balance out one's own body in movement. With balance being placed in the link, the body need not continuously rebalance itself, need not firmly step on the sand.

Design implies emancipation from the grip of a rigid reality. It implies balance in motion—centering. I view such a "floating," self-balancing system of design processes, with no firm ground to step on, in analogy to climbing a sand dune—man linking hands with reality and with his appreciated world—not quite as grim a task as Baron Münchhausen faced in his static trap, the swamp. Also, in movement, evolution will aid us. It even aids physical systems to climb to higher states of organization.

However, linking hands remains important, even vital. By nature, man's consciousness and his appreciated world are closely linked together; often they may even be said to be in a suffocating clinch. But what I mean primarily, is the establishment of a closer link between consciousness and reality—"tuning in" to a dynamic reality which is not autonomous but is interdependent with man in a physical, social, and, above all, spiritual or evolutionary way. In Chapter 9 I shall briefly survey various approaches toward the growth of consciousness through "tuning in," approaches which are spreading quickly, primarily among the young people in America.

There is still another way to represent the design process and to illuminate aspects of balancing, as they relate to human systems. This representation cuts across the relations depicted in Figures 9 and 10 in a new way. If we assume an evolutionary role of man and his systems—and only then—we may legitimately arrange the relations between man and his total space in a hierarchical order, descending from spiritual over social to physical space. In this way, the hierarchical principle represents not a static order but an orientation toward a purpose, the organization of a physical world over a social one to a spiritual world—the *telos* of evolution.

In Figure 13, I sketched such a hierarchical representation in which I equated physical, social, and spiritual space with the same specific forms of human systems which I labeled interference systems in Chapter 4. The relations from Figure 10 would appear here as processes giving rise to ephemeral and changing structures that correspond to different stages of human organization; in ascending order I call them resources (including human, material, and non-

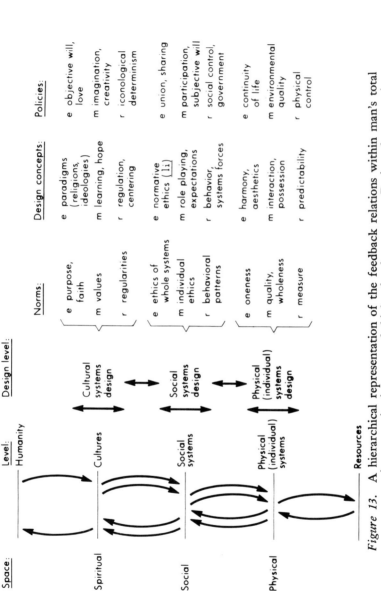

Figure 13. A hierarchical representation of the feedback relations within man's total space, expressing organization toward higher-level purposes. Design does not just "build" a systems level but anchors and balances it between the two adjacent levels. The norms, design concepts, and policies from Figure 10 appear here in their hierarchical relationship (e = notions emerging at the evolutionary level of perception/inquiry; m = at the mythological level; r = at the rational level).

material resources), individual systems, social systems, cultural systems, and humanity. The predominance of processes again implies that none of these structures should be imagined as definitive or rigid—not even the "top structure" of humanity, which is viewed here as the leading force of a dynamic evolution on earth. Humanity, too, is changing in noogenesis.

The relations, manifesting themselves in processes, do not simply translate from one structural level to another. They are viewed here ——and this is of great significance for regulation—as balancing out, as *centering* a particular structural level between both adjacent levels, the one above and the one below. Thus, processes overlap, leaving abundant possibility for redundancy; in Figure 13, I indicated this graphically, unlike Figures 9 and 10. Again, the structures are not given or created outside and independent of these processes, they are defined and brought to life by them. Not the structures regulate the processes—the processes regulate the structures.

The interdependence, the *creative tension between structure and process* reveals one of the most fundamental pairs of opposites from which all life, not just the life of human beings and systems, receives energy. It is perhaps of equal importance as is the tension between form and content in a static view. No resolution of this tension to either side is possible in life—the quenching of processes of change to ensure the solidity of structures leads to suffocation of life, and ephemeralization of structures to let processes roam in complete freedom, will ultimately prevent these processes from touching and building anything. Effectiveness in the bringing about of change may always advance at the expense of rate of change, and vice versa—another manifestation of a generalized Heisenberg uncertainty principle. As in all dynamic phenomena, the real issue is centering.

In Chapter 12 I shall discuss planning in the framework of the systems approach, or more precisely the aspect of vertical centering, in the same terms and with the help of an analogous hierarchical representation in which overlapping planning processes define and center the levels of targets, goals, and objectives. Apparently all evolutionary movements toward higher forms of organization may be represented in such a way—whereas entropic movements may be represented by the familiar organization pyramid, controlled from the top, with information flowing upward between rigid levels, and orders and controls being handed down from level to level. The hierarchical multiechelon system structure (Mesarović, et al., 1970) represents a viable organization form because the units at each level are alive, changing, defined and regulated in interaction with both the next higher level (offering contributions to a higher-level goal, and being coordinated—not controlled—from top) and the next lower level (depending in its goals on the contributions and goals of the units at the lower level).

The first attempts toward building a new organization theory accordingly emphasize the design of systems of processes, rather than structures—the design of inquiring systems (Churchman, 1972) and of horizontal organization systems (Cleveland, 1972). The latter are defined as systems of which no single man can possibly gain an overview and which, therefore, depend on the development of manifold horizontal consultation processes, even if the power of decision is still placed in specific persons. The same trend toward flexibility in the generation of variety, toward horizontal mobility, becomes manifest in many industrial corporations which abandon the vertical product- , skill- , technology- , or resource-based forms of organization and introduce approaches toward organizing around broad social functional categories (Jantsch, 1972).

With processes coming into the foreground of attention, the human actors who initiate and carry them acquire prime importance, not the anonymous structures in which they are administered in social systems. The processes I am talking of here are nothing else than the lives of people and aspects of these lives important to the people who live them. This is of particular significance in the lives of social systems. The processes at work in them are not just the processes emanating from social man, from his specific position in a social structure or from his functions or skills—more importantly, they arise from man's spiritual space, from his interests and ambitions and, above all, from the responsibilities he accepts. Mechanistic organization theory deals with the interplay between functions and skills, and behavioral organization theory with the interplay between statistically correlated more complex group phenomena. But an organization theory based on action deals with real people in a social context—people who bring into play the feedback relationships they maintain within the physical, social, and spiritual aspects of their integral existence.

Thus, at least at the mythological level of inquiry, it becomes of interest who proposes, initiates, or carries a process—who is the *total* man doing this, what values does he hold, what hopes and expectations does he nourish, what myths does he believe in, what is part of his consciousness and what not, what experiences did he gain and what is his view of the world? It becomes of particular interest to find out whether this character is capable and motivated in the direction of innovation, of broadening the spectrum of options open to a social system for its future course of action and thus of giving social evolution a new chance. It may be impossible to judge what action will reinforce an ultimately "winning" mutation of evolution—but it is quite possible to judge whether action enriches or impoverishes the chances for evolution. By this measure we may, in a preliminary way, judge the evolutionary merit of all the processes crisscrossing the physical, social, and spiritual spaces of human systems. But this is only one side of the coin. The *invo*lu*tionary merit is by far the more difficult to assess because we are lacking clear criteria for it.

Figures 9a and 9b (and the latter's elaboration in Figure 10) depict the total design process according to my model, and so does Figure 13, which is a combination of them. The model consists of a total of nine "elements" and eighteen pairs of principal feedback links beween them. All of these eighteen pairs of feedback links are run through not smoothly, but in a sort of vibration, small steps prancing back and forth—indeed, not an easy system to regulate. No wonder that parts of this system always get blocked. If only one of the nine "elements" of the system is blocked, four of the eighteen pairs of feedback links are blocked, too. If the whole spiritual space is blocked—the usual situation of cultural immobility—not less than half of the feedback links cannot work. Cultural and social immobility combined leaves just three pairs of links to operate in the physical space—"the environment" as we like to call it today.

Any one of the representations of the model according to Figures 9a, 9b and 13 may be adopted as an ordering principle for the rest of this book. However, I believe that perhaps the most penetrating view of the processes and tasks involved can be gained by focusing the three following parts of the book on the regulation of consciousness, of the appreciated world, and of reality. There, the basic notions from Figure 10 will reappear. As I pointed out in the beginning of this chapter, such a presentation involves the discussion of feedback processes not only in three spaces (physical, social, and spiritual), but also to both sides: consciousness linked to the appreciated world and to reality, the appreciated world linked to consciousness and reality, and reality linked to consciousness and the appreciated world. However, I shall not formally follow such a partly redundant procedure but deal more freely with the individual feedback links, discussing overlapping links under this or that heading, emphasizing some in detail and grouping others together as I see fit for the purposes of the book. I hope the reader will trust me as I guide him through this study.

PART III

Consciousness Evolving

God guard me from those thoughts men think
In the mind alone;
He that sings a lasting song
Thinks in a marrow-bone.

William Butler Yeats

CONSCIOUSNESS evolves in interaction with reality and the appreciated world. The link to reality is characterized by receptivity to experience and creativity in forming a "represented context of human existence." This learning process can take two ways: an "outer way" of encountering resistance in the surrounding physical, social, and spiritual world and reflecting on it; and an "inner way" of experiencing predominantly nonreflective knowledge, or direct insight, for which a number of techniques have been developed. It is the latter approach which brings into play the human faculty of grasping dynamic reality in a holistic way—of becoming sensitive to the evolutionary aspect of human existence. By centering his consciousness at three levels—which may be labeled heaven, the human world, and earth—man develops a sense of direction for his own life and the life of human systems in the context of an all-embracing evolutionary order of process.

(8)

CONSCIOUSNESS AND REALITY: A DARWINIAN PERSPECTIVE

Consciousness evolves in interaction with both reality and the appreciated world. In this chapter, I shall focus on the feedback interaction between consciousness and reality which is characterized by the experience of resistance—of a hardness from which we escape into the appreciated world which is composed of thoughts:

> Eng ist die Welt, und das Gehirn ist weit.
> Leicht beieinander wohnen die Gedanken,
> Doch hart im Raume stossen sich die Sachen.
> (The world is narrow and the mind is wide.
> Thoughts live easily together,
> But in space hard things collide.)
> (Schiller, *Wallenstein*)

In the three spaces of man's existence we encounter the resistance of matter in physical, the resistance of other persons and of human systems in social, and the resistance of order in spiritual reality. Out of this process of encountering resistance grows a context, a certain holistic view of that part of the world which we have faced during the continuous encounter. Relying on Vickers (1970), I call it the *represented context of human existence*.

In physical space, this process is mediated by the *senses* which transform material processes into psychic processes. A modern view recognize the senses as a field of tension between material and psychic processes (Gosztonyi, 1972). Sensory stimuli become organized into more complex multisensory feelings which, in turn, become organized into even more complex qualities. This process acts by association and induction through the enhancement of other feelings, so that whole systems of feelings and qualities are built, not just an array of isolated elements. The feedback nature of our sensory encounter with the world is borne out by the fact that children up

to the age of four or five have no notion of perspective; they have
to learn it by interaction with the environment. Also, children in
their first years of life have no sense of time; it is learned through
the experience of periodicity in the body. The same kind of process
is at work in social space where it is mediated by *mutual- and self-
expectations* (Vickers, 1970 and 1973). As pointed out in the last
chapter, interaction is strong in both directions in social space—we
may call such a type of interaction communication in the common
sense; all other forms of interaction between consciousness and
reality may also be called communicative but are not equally strong
in both directions. In spiritual space, the feedback process which is
giving shape to a represented context finds a mediator which may
be called *insight*.

TABLE 2
*The evolving norms in the feedback relations
between consciousness and reality. They are
expressions of the physical, social, and spiritual
order encountered.*

	RATIONAL APPROACH	MYTHOLOGICAL APPROACH	EVOLUTIONARY APPROACH
Spiritual Space	Laws (Regularities)	Values	Purpose (Telos)
Social Space	Behavioral Patterns	Individual Ethics	Ethics of Whole Systems
Physical Space	Measure	Gestalt (Quality)	Oneness

These processes are of such great importance for our general
theme because the norms which they evolve in their self-balancing,
and which are the innate norms of man's relating to his world, provide the standards for actively dealing with the world. Any kind of
regulation is only possible on the basis of effective standards, or
norms. Whereas alternatives for active regulation are developed and
tested in the mind in the feedback loops around the appreciated
world, the standards to be used in this sector of the overall design
process evolve out of man's direct interaction with reality—or else
they would be only products of a fantasy dealing with illusory

Utopias. This point is of the greatest significance for understanding man's real options and possibilities.

The explicit norms which evolve from the feedback relationships between consciousness and reality depend on the approach which is applied to express this relationship. In Chapter 5, I have characterized the rational, mythological, and evolutionary approaches as basic modes of inquiry. Table 2 lists the basic norms which evolve in our encounter with reality in the perspective of these approaches; I had already included them in Figure 10.

In physical space, *measure* evolves as the norm of the rational approach. It is significant that measure is part of basic learning only in relatively advanced cultures. Whereas primitive man lived in a world without measure and was initiated into reality through rites which made him share mythically in the lives of the gods (we may also say man was initiated not into the norms relating to physical but directly into the norms relating to spiritual reality), learning became a process of measurement in the classical Greek culture. There, man "measured the world with his body and discovered that the world was made to the measure of man." (Illich, 1970 a). More recently, contemporary man has come to learn "that he is measured by the same scale which can also be applied to things." We may also say, the place of man as a measure for the world has been taken over by technology, whereby not only the familiar scales have become vastly extended, but quite new scales have been introduced as well.

It is not only the scale but the overestimation of the very concept of measurement which is put in doubt today in the context of learning. For Illich (1970 b) again, "men and women who have been schooled down to size let unmeasured experience slip out of their hands. . . . They do not have to be robbed of their creativity. Under instruction, they have unlearned to 'do' their thing or 'be' themselves, and value only what has been made or could be made." The resulting fragmentation of experience has already greatly impoverished human existence and its context. We may even speak of a linearity in the development of experience along a limited number of tracks, the same linearity which has trapped us into patterns of economic and technological growth.

Nevertheless, measure is an important if not all-encompassing aspect of design and, depending on what we mean by design, it will be either human measure or the measure of things. In the latter case, consequently, we ought to speak of the design of technological systems *to which man is fitted*, whether the subject of design is a plastic chair, a computer, a moon rocket, or a modern city of the type which may be found in America but also more recently in Germany and in France (the nightmarish "La Défense" quarters at the edge of Paris or some satellite towns which are designed for a quality of life which is so amiably criticized and ridiculed in Jacques Tati's films). We are virtually drowning in a technicized environment which conditions us but of which, as Marshall McLuhan pointed out, we are as unaware as the fish is of the water in which he swims.

The norm evolving from the mythological approach in dealing with physical space is *quality* or, more suggestively, that holistic notion of form that expresses its quality and function and that is often referred to as gestalt. It is most interesting that the five classical senses seem to be geared to such a mythological approach, rather than to a rational one. Gosztonyi (1972) distinguishes between three characteristics of senses: reality value, formal evidence value, and existential evidence value. The reality value indicates the degree to which a sense transmits material resistance—it is high only for touch and low for the other four classical senses (sight, hearing, smell, and taste). Thus, our everyday life does not provide us with strong evidence for a rational approach based on reality value. On the other hand, both types of evidence value, relating to form and a "quality of existing," are relatively high for the classical senses. Formal evidence value, indicating the degree of insight into formal relations, is high for seeing and hearing and lower for the others. Existential evidence value, indicating the intensity of the experience of the existence of what a sense mediates—quite independent of the reality value (a dream, for example, having zero reality value, may have high existential evidence value)—is high for all five classical senses.[1] Thus, the senses mediate an experience of the physical world which is characterized by a multiplicity of qualities with which we establish a personal relationship, which we mold into a gestalt which combines subjective and objective features.

In the evolutionary approach, finally, the norm evolving in our encounter with reality, is the *oneness* with evolving reality—with a morphogenetic process of which measure and quality are but passing incidences. The senses mediating the corresponding experience are the same ones by which we feel our bodily existence. Gosztonyi (1972) lists the following body sensations: vibration (in which

---

1. Gosztonyi (1972) makes an exception for hearing, which I do not endorse. Noise, more than anything else, makes me experience the existence of other people and of machines, as does also organized acoustical communication.

external sensory and bodily experience become identical), temperature, equilibrium, gravity, depth, tonus (the tension in parts of the body or the total body), and kinesthetics—the experience of movement; many of them overlap and interact with each other. Their reality value varies, but can be high if they are coupled with touching. However, their existential evidence value is generally very high —only here it is not merely directed at the environment but includes the experience of our own nature and existence. Body sensation does not just link us to the environment but also to our own psyche. This is particularly noticeable with tonus, which expresses a basic vital energy in ourselves, and in autogenous movement, such as mime or dance. Many of the body sensations have some periodicity, for example hunger and forms of autonomous movement, such as heartbeat or breathing; in this way, they lead to the development of a sense of time which, essentially, comes out of our own body as "our own" time. Thus, the space/time form of the evolutionary process in a human perspective emerges from the sensory experience of our own body as it moves in the world and along its own process. In the next chapter I shall discuss some of the techniques which have been developed to enhance this evolutionary experience.

In social space, the norms evolving from a rational approach to social reality are *behavioral patterns*—correlations of statistically evaluated empirical evidence from past behavior, as observed "from outside," allegedly without involvement of the observer. Social indicators are here supposed to provide clear and impartial measurements. Amoral social science is based on such behavioral patterns suggesting a regularity that would permit the prediction of social behavior. But the kind of "objectivation" gained thereby, is not only unrealistic but also not the only one that may be sought.

Individual ethics, evolving as the norm of a mythological approach to linking the self to social reality, is another kind of "objectivation" based on a given order which is not to be touched. It stipulates the norms of individual behavior in a social context without questioning this context and the values by which it has been shaped. Individual ethics is usually understood as an integral part of a religion or an ideology—and thus as a static set of rules for the implementation of the values held by that religion or ideology. Underlying is the assumption that man ought to be given such a set of rules to make him fit into a given social and spiritual environment and thus achieve harmony with the social microcosmos surrounding him. Individual ethics ignores the possibility that human action may also interfere

with the macrosystems of human life. Individual ethics, as prescribed
by the current legal framework in most countries, does not prevent
the management of human systems from getting out of hand; it is
seriously threatening the environment and causing growing inequal-
ity on a global scale.

A more dynamic version of individual ethics may develop along
the lines of proairetic ethics, or ethics based on preference. Dorfles
(1972) argues that proairetic ethics ought to guide design, in particular environmental design, and replace purely rational, economic,
and technical considerations.[1] Proairetic ethics would develop out
of the individual's "affective memorization" of the environment, the
personal links of liking or disliking facets of the environment which
we build in relating to our surrounding world in terms of quality.
The establishment of a preference pattern would also imply a
symmetry-breaking process (as I have discussed it in Chapter 5) and
favor asymmetric design—an idea which Dorfles juxtaposes to the
rationality of "a particularly crystallized situation such as gave birth
to some of the great environmental planning efforts of the last thirty
years, from Brasilia to Chandigarh, from Gropius's plans for Baghdad
to the British and Swedish New Towns." Again, a dynamic view
seems to link up with the idea of nonequilibrium systems.

The norm developing out of an evolutionary approach to social
reality is the *ethics of whole systems*, a notion about which Churchman (1968 a) has written extensively. What is considered good for
a human system, may be vastly different from what is considered
good for its isolated members. Some rules may even reverse their
thrust. Under certain conditions, the "Thou shalt not kill" of individual ethics becomes the powerfully enforced obligation to kill
under some stipulated ethics of the fatherland. But nothing is in a
foggier state today than the ethics of whole systems. We still value
managers of countries and organizations by the power of persuasion
they enact upon the members of their respective systems and
systems with which they interact. The mythological approach sees
in individual heroism the self-elevation of man toward the gods, man
becoming a mythological figure himself in his epical struggle. Contrary to this view, the evolutionary approach is interested in norms
which make whole human systems tune in with, and contribute to,
evolution. A good course for a human system would then, for

1. According to Dorfles, an attempt by Wright (1963) to formalize the
proairetic factor in terms of symbolic logic, succeeded in formulating the
problem correctly but was bound to fail in its rational approach.

example, be one along which its capability for such a contribution increases or is at least maintained. It is not easy to measure systems improvement under any maxim—and the evolutionary maxim is perhaps the most difficult to translate into changes of systems parameters.

However, we witness today new attempts to formulate at least part of such ethics in the face of a mounting environmental crisis. Antipollution laws translate some aspects of ethics of whole systems into individual ethics. Damage to public goods enter the cost calculations for private goods. And the new type of human systems which came into being with a new life style, communes, already pays considerable attention to ethics geared to the life of the whole system rather than of its individual members. The boundary line between individual and whole systems ethics is not easy to draw, since both are ultimately expressed in rules addresses to the individual, "Thou shalt" and "Thou shalt not." But what matters foremost is at what level the measure of improvement is applied, at the level of the individual or the system. This is the equivalent of asking at what level a policy is designed. The Ten Commandments, addressed to individuals, are of course geared to the preservation of some social order—but they leave the design of policies to the level of individual lives, and constitute, at the level of human systems, devices of "fencing in" the free space for such individual designs. Thus they focus on individual ethics, assuming at the same time some macroautomatism of regulation at the systems level, once the individual policies stay within a given space.

The distinction to be made is really not just between individual and whole systems ethics but between different levels and sizes of human systems—from the individual, with his individual system of relations and interests, over relatively small groups of people all the way to nations, and further to mankind. It is quite natural that the growing concern about systems ethics leads to new kinds of regimes first in relatively small and well-definable human systems. New ground is first broken in the inconspicuous forms of communal and tribal living. Small human systems are arbitrarily closed off from the bigger systems of society in order to make social experimentation possible at all and avoid the frustrations which large systems hold for similar attempts. However, the experience of syntony in such small human systems is deceptive when applied to larger systems, at least in a time of cultural transition. In larger systems, it is much more difficult to devise systems ethics in such a way that most of the individual interests can be served at the same time.

In a full evolutionary perspective, the ethics of whole systems would not only focus on the melioration of the human systems but of mankind in general—and, ultimately, on the melioration of evolution transcending the survival of any species, including the human species. A belief in evolution implies that man, and all mankind, share in a process of which the evolution of mankind is but a temporary aspect. Such a free attitude, going beyond any particular interests of the individual, system, or even the species, becomes possible if man is also understood as sharing in a universal mind, of being immortal as spiritual energy—whatever the material or nonmaterial manifestation of this energy. In other words, the ethics of whole systems become possible only if we overcome fear—that same fear which is nourished by individualism and the concern with a process which is considered unique and terminates when the individual ceases to exist in its present form. The same fear of termination, of death, is at work in individuals, systems, and the whole human species. The approaches of "the inner way," which I shall summarize in the next chapter, focus on this general theme of sharing in a universal process in which nothing gets lost.

In the spiritual realm, finally, we encounter the resistance of *order* in reality. In the rational approach, we learn about laws or regularities which make the dynamics of the world predictable. These are assumed to be impartial laws, imposed upon the world by an unemotional god who is far removed from us. They form an aesthetic order which we encounter here, an order of the kind which has found aesthetic expression in the mathematical Greek cosmos and the numerical knowledge systems of the Pythagoreans. Such order is also found in Raimundus Lullus's *ars magna* which in the thirteenth century tried to conceive a system of figures and bodies in which all branches of knowledge could be represented, in the mathematicization of notions in Leibniz's *calculus ratiocinator*, and in the mechanical concepts of order which appear in Kepler's *harmonices mundi*, and with Descartes.

The spiritual norms which evolve from a mythological approach to reality, are *values*, conceptions of the principles and enduring themes governing personal relations and processes between ourselves and reality.[1] The notion of value, adopted here, is at a more instru-

---

1. Older definitions of "value" emphasize its "desirable" character. However, such a definition is begging the question of how we conceive something as

mental level than the classical "primary values"—the true, the good, and the beautiful. The latter may be viewed as the basic themes of order emerging from the evolutionary, the mythological, and the rational approach, respectively. The traditional hierarchical order of these primary values, with the true on top, also points to the hierarchical principle applicable to the three approaches.

The values of which I speak, appear as systems of interconnected values, forming a moral order.[1] Each individual, social, or cultural consciousness holds such a system of values which is not fixed over a lifetime but fluctuates according to the shades the various components assume and to their interrelations. Values are not clearcut and unambiguous conceptions, but may be viewed as tension fields between pairs of opposites, as interference patterns of complex wave functions. The opposites are always present in us, although we often attempt to resolve them by choosing one or the other. But the full quality of a value emerges only from this tension between opposites, which has to be lived out. We find here again the complementarity of the yin/yang principle, with each side containing the seed of its opposite. Value pairs that have come to the foreground more recently, making the tension between them felt more acutely, include for example:

Competition	vs.	Responsibility
Growth	vs.	Sufficiency
Efficiency	vs.	Effectiveness
Power	vs.	Equity
Entrepreneurship	vs.	Security
	etc.	

Much of the challenge of centering arises from such tensions in value dimensions that move into an "acute" phase.

Values are not invented or imagined. They form part of the basic bondage between ourselves and reality. Depending on the nature and emphasis of our interaction with reality, encountering the resistance

---

desirable or not. In my model, I introduce the desirable in the interaction between consciousness and the appreciated world.

1. Hierarchical or priority-ordering schemes of values—in particular Maslow's popular five priority levels: physiological needs, safety (security), belonging, esteem, and self-fulfillment—assume a static world of the "here and now," in which values can actually be satisfied in sequential order and "shelved away." As if any of the mentioned categories could ever be satisfied and put aside! Their relations within a system of values may, in certain situations, emphasize the influence of a particular value category—but always in the presence and under the influence of others.

of moral order, some value dimensions may emerge very promi-
nently and others may remain in the subconscious. Like our genetic
potential, the value system is only partially in play at a given
moment. Not only is a continuous fluctuation introduced by the
tension between pairs of opposite values, but also changing relationships between such pairs and clusters of pairs. An alive value system
in the consciousness of individuals, organizations, institutions, cultures, or the whole world is the prerequisite for flexibility in attitude
not just toward reality but—even more importantly—also toward
the appreciated world and thus toward the human creative potential
and its application. A flexible value system invalidates the Aristotelian logic in the domain of qualities. Without the notion of a value
system, it would, indeed, be highly puzzling to find that a person
prefers, among a choice of products or services or entertainments,
A to B and B to C, but—against all Aristotelian logic—C to A. But
such is the play of attitudes in a mythological world, in the world
of qualities in which we live with our emotions and idiosyncrasies.
No wonder that rational approaches fail so miserably.

Value systems may also be understood by analogy to highly nonequilibrium, dissipative structures. The more conscious we become
of our lives, the more we experience the fluctuation between
opposites and in the relations within the system. In our time, we
are becoming painfully aware of the tensions between materialistic
and nonmaterialistic values, both of which we would like to maintain. People's value systems are in dissonance with each other and
even in open conflict. The value systems held by the declining
culture, as well as individual value systems, have entered a phase of
increasing instability—announcing an impending evolutionary jump,
as has perhaps already occurred with that part of the youth which
has changed to a simple, self-sufficient, inoffensive, and essential life.
The spiritual renaissance among the young generation, particularly
in America, is an impressive manifestation of such an evolutionary
jump beyond all education and social incentive. It even turned
Maslow's above-mentioned "logical" priority scheme upside down—
self-fulfillment comes out strong as a governing and coordinating
category.

The order which is learned through an evolutionary approach,
finally, is *purpose* or that sense of direction for which the British
Prime Minister referred to the archbishops. Purpose is meaning in a
dynamic perspective and forms indeed the core of any religion or
evolutionary myth. But the sense of purpose which evolves from

interaction with reality relies on a kind of purpose which evolves from interaction with reality, relies on a kind of direct insight that is not developed in the laboratory of the appreciated world. It is the direct experience of an objective purpose which may be sensed by "tuning in" to the evolutionary movement. In the following chapter, I shall discuss several approaches which seem to hold some promise in this respect. But before, I should like to mention two more aspects of the processes which I have outlined in this chapter.

The first aspect concerns the correspondence between the macro- and the microworld in this encounter with reality. "As above, so below; as below, so above"—the principle of Hermetic philosophy has been a recurrent theme through the millenia of history of human thought. I shall return to this important principle in the following chapter. Here, I want merely to point out that only such a correspondence between above and below permits use of the norms evolving from our interaction with reality in a more general context to gain a world view. The moral order which emerges from a mythological approach is also assumed to be the moral order of the gods—by valuing the good in man, we expect at the same time to value the good held by the gods. In ancient China, this universal moral order underlying evolution was known as *li*. Confucius made it the basis of social regulation:

Guide the people with governmental measures and control or regulate them by the threat of punishment, and the people will try to keep out of jail but will have no sense of honor or shame. Guide the people by virtue and control or regulate them by *li*, and the people will have a sense of honor and respect.

The modern notion of an *ethics of whole systems*, which I have discussed above, now appears in a new light: it is not uniquely and *ad hoc* defined for individual systems but falls together—or, more precisely, is hierarchically structured by—the universal moral order of *li*. Therefore, in Table 2, the ethics of whole systems is correctly listed as emerging from an evolutionary approach to inquiry.

The purpose which emanates from an evolutionary approach is at the same time taken to be the purpose of all evolution and of individual man who has an active role in it. The principles of evolution seem, indeed, to be the same at all levels; communication processes, structures, and mechanisms of evolutionary mutations in dissipative structures or topological modes of change seem to apply uniformly throughout all evolutionary processes. In a rational approach, the inductive process has been reversed—what holds in the physical universe, with its impartial laws of motion, has been transplanted to the world of humans.

The second aspect I should like to emphasize and at which I have already hinted in Chapter 5, concerns the mismatch between perceiving and dealing with the world, which is at the root of many of our current calamities. A look at Table 2 (see p. 132) makes it clear that we live primarily in a mythological world whose values we strive to pursue, whose individual ethics we accept and whose qualities we relate to with our whole person. But the intellectual instrumentarium which we apply in dealing with our mythological world more efficiently, belongs to the rational approach, from Newtonian mechanics to behavioral science, from economics to operational planning, from logic to system analysis. We try to quantify our world of qualities only to find that its quality becomes elusive. The systems approach, as I shall briefly discuss it in Chapter 12, constitutes a first attempt to bridge the gap between a rational and a mythological world. On the other side, in a mythological world we also lack a sense of direction and purpose—we relate in an existential way to a spatial cross section of the world, not to its order of processes.

It becomes dramatically clear that in dealing with reality, we have to take all three approaches simultaneously. Such an integrated approach might then look as follows:

—An evolutionary approach provides a sense of purpose and direction—the "know-where-to";
—A mythological approach provides a sense of quality and systemic relation—the "know-what";
—A rational approach provides a sense of operational efficiency and determinacy—the "know-how."

All three approaches together only provide a meaningful basis for human design. The full-scale design process has to integrate all three of them.

In encountering resistance through interaction between ourselves and reality, our consciousness—or the consciousness of any social or cultural group and organization—acquires primarily the experience of what we *are*, what we *feel*, what we *perceive*, and what we *know*. Interaction between the self and the appreciated world will produce other basic experiences, namely the experience of what we want, conceive, and can do. These additional processes of shaping consciousness will be discussed in Chapter 10, and subsequently the challenge of centering our consciousness in the flux of these experiences.

Before this, however, I should like to outline in the next chapter some approaches and techniques for an "inner way" of interacting with reality. If, in line with evolutionary terminology, we call the formative experience of encountering resistance in the world the external factors in the evolution of consciousness, the "inner way" would then correspond to the internal factors. If we cannot hope to bring such internal factors into play effectively, we shall be lost in our mythological world and in its system without a sense of direction. If we cannot develop approaches to enhance these internal factors and call upon them when we need them, our design process will be blind. In a time of cultural transition, such a sense of direction becomes of foremost importance for any reasonable kind of design.

INTERNAL FACTORS IN THE EVOLUTION OF CONSCIOUSNESS

The processes of interaction between consciousness and reality which I described in the last chapter build man's relations to his world primarily at the rational and mythological levels of inquiry. They mainly permit a kind of static orientation. Where a sense of movement does enter, it refers to the movement of self in a more or less static world rather than to a sense of overall movement, a sense of universal evolution. Yet it is this latter sense which is needed if man wants to act in harmony with evolution or, beyond that, wants to take seriously his role as an instrument and, indeed, a "cutting edge" of evolution.

At the rational level, man looks at a creation which has developed outside of himself. At the mythological level, by means of self-reflection, he tries to fit into it, to define his very own niche which permits him to build his anthropomorphic world. But at the evolutionary level man is an integral part of evolution, he feels the dynamic forces of the universe acting through himself as a channel for them. In Chapter 5, I employed the image that, at this level, man is no longer steering in the stream of evolution, he *is* the stream. Access to this level is generally gained through a *nonreflective* attitude paired with a particular kind of sensitivity and openness. Whereas at the mythological level it is primarily external factors of evolution that come into focus—human consciousness and the self being shaped by the Darwinian processes of coping with outer resistance and matching the requirements encountered in reality—internal factors in the evolution of individual consciousness seem to dominate at the evolutionary level. They have been recognized under different names, such as "man's metaphysical urge," or "the Zeitgeist," or Kant's "categorical imperative."

These internal factors are not easy to grasp explicitly. They are accessible only through an "inner way" which I propose to discuss in this chapter, having dealt with the "outer way" in the last chapter. "The way inwards," explains Dürckheim (1956), "rests on three factors. The first is an *experience* wherein the light of Being illuminates the darkness of life. The second is *insight* into the relationship of his worldly I and his transcendental Being, as well as into the difference between the state in which he is cut off from his Being and the right state which opens him to it. The third is *practice, exercitium*, which corrects the wrong working of the misguided I and builds up the right attitude in a right way. That is a right attitude in which a man is permeable to the Greater Life which he embodies and by which he is enabled to perceive it in the world. He is then truly himself and the world of space and time becomes transparent for him in the Being which transcends space and time. The way leading to this condition is by the transformation of the *whole* man, i.e., a unit of body, mind and soul."

The experience of which Dürckheim speaks, may be triggered by any encounter with reality, also in pursuit of an "outer way." The insight reveals internal factors at work at the evolutionary level. The *exercitium* makes use of a body of approaches and techniques which I shall summarize later in this chapter.

The internal factors in evolving consciousness may be considered a valid source for obtaining a sense of direction for tao, the right path in the world, if *one* assumption is granted which I have already evoked in this book and which has been made over and over again in the spiritual history of mankind. This is the assumption that the evolution of human consciousness is an image of the evolution of the universe, or in A. Huxley's (1954) words that "each one of us is potentially Mind-at-Large." This linkage of the dynamic order of the human world with that of the whole universe, of micro- and macroscopic aspects of evolution, is also referred to as the *law of correspondence*. It has formed the core of cosmology in various versions from Hermetic philosophy through Plato and all the way to medieval alchemy and to the rationalists of the seventeenth century, where it found expression in Leibniz's monadology: "Consequently, each body feels all that passes in the universe. . . ." It is and has been the basic assumption on which all mythology and all forms of religion rest. The New Testament, giving meaning to God's life in the human flesh, is just one expression of such linkage of heaven and earth, expressed in a dualistic spirit separating the two domains. Shiva's dance, the divine erotic penetration of the earth in Hindu religion, is another one, wider and all-embracing, resolving any dualism; man becomes separated from God only to the extent that he becomes uncentered. In Hindu mysticism the law of correspondence is expressed in the identity of *atman* (the true essence of self) and *brahman* (the true essence of reality); thus the search into oneself becomes at the same time the search into outside reality. This is the spirit of the "inner way."

In reaction to the dualism which haunted Gnostic thought through the centuries, Jonas (1969) has perhaps given the most comprehensive expression to the ultimate union between heaven, universe, earth, and man in the great thought that God has given himself up in order to flow into the evolution of the universe. Successive evolutions of universes would then be transformations of God—transformations which are predetermined in their general waves but uncertain and risky in their outcomes due to the element of free will acting in an evolution through self-reflecting, intelligent beings such as humans.

In a hierarchial perspective, this "downward" acting principle seems to match the "upward" acting principle—expressed in many intuitive world views and already mentioned in Chapter 6—that consciousness is structured as an *infinite regression of levels* (physical reality consciousness within a higher consciousness within a yet higher consciousness, and so forth, ad infinitum). Proceeding along this regression opens up knowledge from the individual over the collective and the cultural level to the level of mankind and, finally, to the working of an overall universal evolution, of energy unfolding, and to the void (the "All") beyond the evolution, a memory of which seems to express itself in our deep longing for peace. All this knowledge, and this is the crucial point, is available *within* ourselves.

The structure of brain-mind processes constitutes one of the great fields of ignorance in science that has come to the foreground of interest recently; the others are the form of a unified theory of physical field-particles, and the character of biological organization (Whyte, 1965). All three are aspects of an overall process of evolution which cuts through all forms in which energy becomes manifest—physical energy/matter, life, and spiritual energy/consciousness.

As the DNA, the molecular genetic material, of each cell contains all information relating to a complete organism, so may the total memory be viewed as being in all parts of the brain. (Pribram, 1969). Recent research has produced certain hints which may soon become substantiated and synthesized into a theory of holographic brain function along the lines suggested by Pribram; in a wider framework, the brothers McKenna (1972) have suggested the same principle for mind quite generally. In such a theory the mind, like DNA, may be assumed to be linked to the totality of information available: "If each mind is a holographic medium, then each is contiguous with every other, because of the ubiquitous distribution of information in holography. Each individual mind would thus be a representation of the 'essence' of reality, but the details could not be resolved until the fragments of the collective hologram were joined." (T. and D. McKenna, 1972).[1] This also would provide a rationale for the enhanced capability of groups of people to "tune in" and focus on new insights, as well as suggest a mechanism by which syntony may develop among people as a collective process rather than through individual persuasion.

If each of us thus is Mind-at-Large, to repeat Huxley's formulation, the ultimate task for human creativity is not to invent and create *ab novo*, but *to find*. Bergson already has stated that the function of the brain and the nervous system is eliminative, not productive. The right attitude may then be compared to that of a hunter or fisherman—usually, one does not go out to hunt one particular stag, or catch one particular fish, but one goes into the appropriate preparations to ensure maximum likelihood of hunting down stags or catching fish in general. "Be the spider, not the fly," is also one of the advices given by Buddhism. If creativity is primarily finding, a holographic theory of mind would also provide a basis for the definition and explanation of *social* (or *collective*) *creativity*. Groups of people would then be capable of finding more information, of becoming more creative, in an evolutionary context.[2]

1. The physical phenomenon of holography provides a suggestive analogy: If a holographic plate (showing a pattern of interference images) is illuminated by coherent light, the three-dimensional image of the photographed object appears in space (as the manifestation of a standing wave-form). If the holographic plate is cut in half, or into a number of small chips, the illumination of each chip yields the *total* three-dimensional object, but with less resolution of details.

2. If creativity is understood as the receiving of messages, an analogy may also be seen here to the receptivity for weak natural electromagnetic fields.

Following such lines of thought, which may still look to be somewhat on the speculative side at the present moment, information would then appear as standing wave-forms resulting from the interference of mind waves, of processes. We find here the same basic idea I have tried to incorporate in my model of the human design process—consciousness, appreciated world, and reality (even physical reality, as modern physics bears out) all may be best understood as interference patterns, as standing wave-forms resulting from the interaction of processes. The McKennas (1972) would see mind moving freely on a four-dimensional hyperplane—the four-dimensional reality of space and time, as it has been introduced by the theory of relativity and characterizes the evolutionary level in my scheme. Reality, and indeed the whole universal evolution, would be represented by a four-dimensional wave,[1] but the individual person, through mind, and utilizing energy in correct ways, would be free to go to the places and think the thoughts that are close to the appropriate tao, the right way, within the givens of one's historical position as constellated by the basic wave-form. The latter may be described "by an energy map of changes running forward and backward upon themselves, a sort of alternating current." (T. and D. McKenna, 1972). This alternating current provides for a maximum degree of freedom in the human response. Using the image of the sailing boat from Chapter 5, each direction of the current may be compared to a strong wind blowing in one particular direction and the human response to the flexible operation of a sailing boat. The old dilemma of determinism vs. free will appears here in a totally new light and virtually disappears in a multi-level view of the same reality, which seems to come very close to a view at the rational and mythological levels in the sense I use these terms.

In the entire domain of life, this receptivity has been found to be the higher the more highly organized the receiving system is. Biological systems react more strongly to fine vibrations than organic systems which, in turn, react more strongly than cellular systems which, in turn, receive finer vibrations than molecular systems (Presman, 1970). Perhaps the optical holographic and the electromagnetic image ought to be viewed as complementary aspects of the basic creative process.

1. The McKennas (1972) believe they have identified its form in the sequence of sixty-four hexagrams forming the *I Ching*, and link it to the sixty-four codons of DNA, thus for the first time suggesting a scientific first principle for an evolutionary cosmology, embracing both the biological and the physical world and their phenomena.

Whether the above suggestions anticipating a forthcoming theory of mind and evolution are considered acceptable or not at this stage, I should like to propose the following minimal notions which provide a constructive basis for the discussion of a variety of approaches to "tune in" and obtain a sense of direction:[1]

—Unity of macro- and microaspects of mind and evolution (the law of correspondence).

—Existence of an evolutionary tao, an objectively right way, for each form of movement, in particular for each manifestation of life; further, the expression of this tao in the "coordinative conditions" (Whyte, 1965) which give rise to internal factors of evolution.

—Unity of the three levels of perception (rational, mythological, evolutionary) in all manifestations of human life, including reflection and self-reflection, so that any deviation from the tao may be sensed and determined at any one level; this assumption also lets appear as valid approaches which focus on one particular level only.

—A communication mechanism which is at work across the three levels of perception, so that, for example, "insight" from the evolutionary level may be received in some other form at the mythological level, e.g., in the form of intuition, or dreams, or general vibrations felt as quality.

All four assumptions are interdependent. A dynamic view of the law of correspondence, in particular, implies a certain structure of the evolutionary tao. This becomes clear in the construction of the *I Ching*, the ancient Chinese Book of Changes. When "in ancient times the holy sages made the Book of Changes," explains the *I Ching* (see *I Ching*, 1967), "they put themselves in accord with tao and its power, and in conformity with this laid down the order of what is right. By thinking through the order of the outer world to the end, and by exploring the law of their nature to the deepest core, they arrived at an understanding of fate." This understanding of fate expresses itself in the structure of tao which becomes the constructive principle for the *I Ching* itself. Tao is to be conceived in terms of three integral aspects, pertaining to three levels—the tao of the earth, the tao of man, and the tao of heaven.

The three taos and the three levels of perception and inquiry seem to correspond in a certain way—they are not identical, but they coincide with their core notions. This may be understood as

---

1. These minimal notions are indeed the ones which need to be accepted if a wide spectrum of phenomena is to be taken into account which cannot be explained by present science, but is accessible to scientific investigation: for example, telepathy, telekinesis, plant communication, healing, acupuncture, psychic reading, and shamanism.

another dynamic aspect of the law of correspondence: man trying to match objective reality (the aspects of tao) and his subjectively attuned powers of inquiry and response. Thus, the following correlation may be drawn:

—At the rational level, the tao of the earth is put into perspective; the focus is on cause/effect relationships in a rigorous or a statistical-probabilistic form, including instinctive behavior.
—At the mythological level, the tao of man is put into perspective; the focus here is on the wide spectrum of expressions of humanness through subjective unfolding and transformation of energy.
—At the evolutionary level, the tao of heaven is put into perspective; the focus for man here is on "tuning in" to the evolutionary wave-form, in other words on developing a consciousness capable of relating to a four-dimensional reality.

The overall sense of direction emanates from the evolutionary level; it constrains and guides man's subjective responses at the mythological level. Thus, the crux of the matter is: Can we learn to "tune in" to the evolution and to interpret the corresponding indications which may reach us at any level? In this chapter, I shall try to give a very brief but systematic overview over old and new approaches, putting into perspective their contributions to man's coping with his oldest and most essential and basic task in life. These approaches all hang together, they form a system. When the whole system becomes actualized through searching by way of several approaches, we may receive information of enhanced value and pertinence. The system hangs together at least in the following principal ways:

—The three levels and the three aspects of tao hang together, as I have already outlined above. This becomes visible, for example, in myth formation, the oldest approach to "tuning in" to the evolution. As Campbell (1972) points out, myth formation receives its impulses from all three levels: mortality and the desire to transcend it, the social order and its endurance, and the universal order. The evolutionary level is generally less directly accessible than the other two levels, but what happens at the evolutionary level is reflected at the other two levels as well.
—Learning, healing, and enlightenment all appear as integral aspects of the same "growth" process (Naranjo, 1972). In this context, it is interesting to note that Michael (1973), in his search for the kind of people who could do planning not as an engineering, but a learning activity, looks with considerable hope to the approaches proposed by humanistic psychology which also emphasize the unity of learning, healing, and enlightenment (or, more modestly, happiness).
—Tao, a right way, and the techniques employed to sense it, hang to-

gether because these techniques also constitute the forms to pursue, to live out tao as it is again best expressed by the *I Ching*: "What is above form is called tao; what is within form is called tool."
—Both dimensions of a genealogical approach—the phylogenetic approach looking back over mankind's evolution, and the ontogenetic approach attempting to grasp an individual's development—fall together. The well-known fundamental biogenetic law, "Ontogeny recapitulates phylogeny," is not only valid in the realm of physical structure, but extends also into the realm of consciousness, or spiritual structure!

In these formative principles, the basic feedback nature of this whole system of approaches already becomes recognizable. "Therefore there is in the Change the Great Primal Beginning," says the *I Ching*. As Naranjo (1972) points out, approaches along a way of action may take the direction of self-expression or of self-restraint; approaches along the ways of feeling may take the direction of devotion (cultivating the most real and basic feelings) or of catharsis (reflecting the tendencies of the moment in all their imperfection); and approaches along the way of knowledge may take the direction of intuitive thinking (the poet who only asks to get his head into heaven) or of discursive thinking (the logician who seeks to get heaven into his head). We find here the old relationship between subjectivity and objectivity which characterizes man's relationship with the world around him and within himself. Taking one preferential direction is worthless. The reality of the "inner world" evolves through the same feedback process which characterizes all aspects of the human design process.

At the core is always the same *re-ligio*, the same linking backward to a primal beginning, to which I have already pointed in various contexts. In fact, the design process, as I described it, is nothing else but such an attempt to "tune in" to evolution—it is learning, healing (improvement), and enlightenment in one and the same process.

I distinguish basically between four lines of approaches or techniques: the *genealogical* approach is the journey backward to the origins in its most direct form, both in its *phylogenetic* and its *ontogenetic* version, considering humanity's or an individual's process, respectively. There is also an *ontological* approach which does not attempt to measure out any movement over time, but attempts to sense the direction "from which the wind is blowing" at a given moment, as if by holding up a wet finger. Finally, there is a forward-oriented *kinesthetic* approach which focuses on acting out movement, and experiencing the "pure essence" of movement.

To the extent that the genealogical and the kinesthetic approaches depend on the actual *observation* (and not just the basic sensing) of a movement, they are subject to uncertainty in analogy to the Heisenberg principle. Understanding well the speed and direction of movement implies considerable uncertainty as to the current position, and vice versa; but it is clearly the sense of movement which is important here, not any clear definition of position or speed. The essence of approaches to direct experience at the evolutionary level, and particularly along the ontological and kinesthetic lines, is a totally open, *nonreflective* attitude. Fischer (1971) has interpreted meditative and ecstatic states (which are the goals of the ontological and kinesthetic approaches) as opposite ends of a continuum of mental states, which include "normal" reflective awareness in the middle. Meditative states mark the peak of progressive trophotropic arousal, ecstatic states the peak of progressive ergotropic arousal. In this scheme, the experience of "oneness" which provides access to the evolutionary level of perception, may be explained by the integration of cortical and subcortical activity in the nervous system—or to be more precise, this is a possible rational aspect of an explanation.

The "inner way" is often discussed in terms of notions which are borrowed from Far Eastern mysticism and philosophy. These notions are indeed helpful since they developed with a practice of the "inner way" as the very focus of human life, much more so than in Western forms of culture.

The four approaches which I have outlined above correspond to the four Hindu *sadhanas*, or ways to enlightenment: the phylogentic approach to the *Way of Knowledge*, or *Jnana Yoga*; the ontogenetic approach to the *Way of Compassion/Love*, or *Bhakti Yoga*; the ontological approach to the *Way of Meditation and Mind Control*; or *Raja Yoga*; and the kinesthetic approach to the *Way of Movement/Action*, or *Karma Yoga*. They emphasize different aspects of the same process, in the same order: collective learning, healing, enlightenment, and individual learning. I call them *ways of integration*.

All four approaches may be characterized by an ascent along the seven *chakras* of Hindu mysticism, delineating an evolutionary path of gradual integration. The seven chakras may be understood as

energy centers of the body, or as levels of consciousness, or—
perhaps most suggestively—as levels of energy exchange, or generally metabolizing activity at which energy is received and dissipated.
In simple terms, the seven chakras may be characterized as follows:

First chakra: Potential for physical survival, animal mentality.
Second chakra: Pleasure and sensual gratification, sex.
Third chakra: Power, mastery over people and environment, "ego trip."
Fourth chakra: Heart, compassion, and love transcending ego, "universal love."
Fifth chakra: Seeking of God, the spirit of the Adagio from Bruckner's Eighth Symphony.
Sixth chakra: Wisdom, the "third eye" on the forehead.
Seventh chakra: Full enlightenment, union with God.

As energy exchange reaches higher chakras, and the evolutionary
path is ascending, the lower chakras become coordinated under the
higher chakras. We may say that the chakra system is a multiechelon system, not a stratified system in which one would live only
at one energy level at a time (as many Western people actually do).
The chakras may also be understood as dynamic regimes of human
consciousness, and ascent toward higher chakras as mutations to new
regimes permitting renewed and fuller energy exchange. The basic
evolutionary process of "order through fluctuation," which I have
evoked in many contexts in this book, may also be assumed to
apply to the growth process along the ladder of chakras. In Chapter
14 I shall also discuss the evolution of mankind generally in terms
of these chakras.

The four *sadhanas*, or ways of integration, provide one ordering
principle for the approaches to an "inner way." The three aspects
of tao, and the corresponding three levels of inquiry, provide another principle—they may be called ways of exploration, or *tattvas*
in the ancient Hindu scheme: At the rational level, the tao of the
earth emerges from an objective way of awareness (or mindfulness);
at the mythological level, the tao of man emerges from a subjective
way of feeling; and at the evolutionary level, the tao of heaven
emerges through an objective way of revelation and/or insight.

The four *sadhanas* and the three *tattvas* yield a matrix of twelve
fields by which specific approaches or techniques may be ordered.
This matrix, depicted in Table 3, may be understood as a simplified

TABLE 3

A systematic overview over approaches and techniques of the "inner way."

	GENEALOGICAL APPROACHES		ONTOLOGICAL APPROACH	KINESTHETIC APPROACH
WAYS OF INTEGRATION (SADHANAS)	PHYLOGENETIC APPROACH	ONTOGENETIC APPROACH		
WAYS OF EXPLORATION (TATTVAS)	*Way of Knowledge* (JNANA YOGA)	*Way of Compassion/ Love* (BHAKTI YOGA)	*Way of Meditation and Mind Control* (RAJA YOGA)	*Way of Movement/ Action* (KARMA YOGA)
EMPHASIS	COLLECTIVE LEARNING	HEALING	ENLIGHTENMENT	INDIVIDUAL LEARNING
Evolutionary Level Tao of Heaven *Way of Revelation/ Insight* (objective)	Cosmology Astrology I Ching Collective Unconscious (Jung) Tao or no-knowledge	Spiritual psycho-synthesis Logotherapy Psychic healing, massage Initiation rites Actualism (Agni Yoga) Kundalini Yoga Acupuncture, Shiatsu massage, etc.	Satoris Shamanism Noh Esalen massage Meditation Za-zen Mudras	Alchemy Rituals, drama Esoteric schools Dervish dancing Tai-chi chuan Japanese tea ceremony

Mythological Level Tao of Man *Way of Feeling* (subjective)	Mythology Cults Archetypes Tarot History : political arts philosophy religions	Psychoanalysis and -therapy Gestalt therapy Psychosynthesis Encounter techniques Psychosomatic therapy	Arts Drugs Tantric exchange (Schizophrenia)	Mime Psychodrama Role playing techniques Theater games Primal therapy
Rational Level Tao of the Earth *Way of Awareness* (Mindfulness) (objective)	Ecology (natural and social)	Grounding Bioenergetics Structural integration and patterning Polarity therapy, reflexology	Transcendental meditation (contemplation) Isolation Fasting Biofeedback Sensory awareness Mentations	Chanting Drumming Gymnastics Movement (dancing), eurhythmics Martial arts (archery) Hatha Yoga Calligraphy Technology

model of the *system of the inner way*, a system which may be activated through any one of its elements or through a combination of them. Activating the system is but a metaphor for man's striving to build—or, as it now seems more correct to say, *find*—an inner anthropomorphic world which can be his home. Some approaches, listed in a specific field of the matrix according to Table 3, are particularly powerful in integrating the three levels, for example the arts and drugs. Since the evolutionary level is not readily accessible in a direct way whenever we need it, vertical integration is the foremost aim.

Mozart's opera *The Magic Flute,* based on Schikaneder's libretto which is so heavily influenced by Freemason symbolism and wisdom, provides perhaps the most beautiful and profound image of ways of integration and of the three aspects of tao, which appear here as prototype levels of human existence: Sarastro and the priests live the tao of heaven, taking the Way of Meditation and Mind Control; Pamina and Tamino find their way to the tao of man through the Way of Compassion/Love (including some psychotherapy to free Pamina from the influence of her possessive mother, and some logotherapy and psychosynthesis for Tamino, and elaborate initiation rites for both); and Papageno, clumsily and involuntarily taking the Way of Movement and, *contre coeur*, learning at least to shut up and obey, wins his good life with Papagena along the sensual and instinct-guided tao of the earth.

The ways of vertical integration across the three levels mean literally the integration of body, soul, and mind—or of Kath (the vital center), Oth (the emotional center), and Path (the intellectual center) as the corresponding energy centers of the body are called in the ancient terminology of some of the contemporary esoteric schools. The Kath is anchored in the tao of the earth, the Path in the tao of heaven, and the Oth lives out the tao of man; thus, the three levels also represent the three energy centers of the human body.

"Sinnlichkeit ist Sittlichkeit" (sensuality implies ethics)—Goethe's insight is revived in the view of the body as "preternatural temple," and the person as a "microcosmos in which universal order is *created*" (Rock, 1973). By "listening to our body," we hold our ears to the forces of evolution as they express themselves through the body. *Tantric Yoga*, also called the Way of Energy, may be seen not just as another way of integration, but as a general methodology for the transmutation of energy leading (along the sequence of the seven chakras) toward increasing union. It may also be called the Way of Self-Attunement or Self-Expression—equating the principle of "Seek the Lord and live" with the principle of "Be yourself." It plays a role in all classic *sadhanas*. Because to listen to ourselves, and in particular to our body, is the only way to approach the divinity in a nondualistic way.

It is evident today in the Western world that the mind center, or Path, has become overdeveloped in comparison with the two other centers. In particular the Kath, the center of instinctual wisdom in the lower belly (the center of gravity), has become almost totally neglected in Western society, although it still makes itself felt through powerful "gut feelings" in times of crises. Its development is the central concern of some of the esoteric schools and "growth centers" in America (in particular, of the Arica Institute). The way of life implicit in the teachings of the Yaqui sorcerer Don Juan (Castaneda, 1969, 1971, and 1972), so influential among young people in America, also is a way of the Kath, a life as a warrior and hunter for natural powers. With Path and Kath thus well taken care of, the development of the Oth, the emotional or love center, seems to be pitifully neglected by either side at present—and yet, as I shall try to argue in some of the following chapters, it will soon become of even more crucial importance than either of the other two, being called upon to carry the emerging task of cultural organization (see Chapter 14).

It is interesting to compare the more recent esoteric schools with older anthroposophian (Rudolph Steiner) schools and their concern with integration. There, the first seven years of schooling focus on the development of the will (which may be seen as corresponding to the Kath), the second seven years on the development of feeling (the Oth), and the last seven-year period only on the development of thinking (the Path).

The large number of specific approaches and techniques which form the *exercitium* of the "inner way"—Table 3 gives only a few representative key words—may appear confusing to anyone taking a first look at them. Naranjo (1972) speaks of one hundred and fifty techniques which he had already considered several years ago in a systematic survey for the Stanford Research Institute, and the number has certainly increased since then. Many of them emerge from a relatively new orientation in psychology called "humanistic psychology"—what humiliation for conventional psychology which thus appears branded as mechanistic!—and are often grouped under headings such as "human potential movement" or "existential technology." This latter name already implies the basic *caveat* which is due in dealing with these approaches: many of them have sprung from the rediscovery of the "here and now"—man becoming sensitized to the forces acting upon his life at a given moment and place—which has led to a vastly exaggerated one-sided concern with the momentous present which appears as an absolute value. The "here and now" along the lines of psychologists such as Maslow and Lowen still provides the reigning slogan for the American west coast which lives its wave of existentialism, epitomizing the potential autonomy of the stationary self, more than a quarter century after Sartre had made his profound impact on Europe from his table at the Café "Flore." But the existentialist attitude inherently refers only to the mythological level of inquiry and perception and, if taken absolute, excludes an attitude of openness toward the dynamics of ontogenetic and cosmic evolution. It may be held responsible for much of the immaturity, the weakness in the acceptance of personal responsibility, the wave of hedonism, the lack of concern for society's future, and the general carelessness of the "I'm o.k.—you're o.k." spirit, which afflicts part of the younger generation at present.

However, it is often the existentialist attitude which is at the beginning of sensitizing people to the evolutionary level, to the dynamics of their lives, and makes them care for their own future and the future of mankind and the planet. The step from a shallow pleasure-oriented existentialism—because it is in the name of sensual pleasure, not "la nausée" (nausea), that California celebrates its version of existentialism—to an evolutionary orientation is apparently quickly taken. In fact, quite a number of people seem to have taken it amidst the wild mixture of primarily existentialist techniques presented at the Esalen Institute in Big Sur, California, the largest and most renowned growth center in America, operating since 1966. Existentialism, in an evolutionary perspective, appears as a form of reductionism which may be overcome if the way of integration does not stop there.

An existential attitude can already serve to help man in breaking out of rigid physical, emotional, and social patterns of life, usually identified with an "ego" which is seen as an encrustation brought about by processes of schooling and socialization. What the "ego" overcrusts, is called man's "essence," his sharing in the overall stream of life, in the divine principle. These are just new words, not new notions. In Jungian terminology, the "essence" corresponds to the core of self, the center of one's being, which also corresponds to the *atman* of Vedanta. Jung called persona what is now called the "ego," and used to be called *ahankara* in Vedanta. The dichotomy between "ego" and "essence" is a recurring theme running through the entire history of the "inner way."

Some schools profess to drive out the "ego" altogether—but this would not leave any possibility for focusing energy, for acting instead of merely behaving. Thus, a more realistic ideal pursued with these techniques is personal *flexibility* in changing one's "ego" patterns, a fluidization of the personality approaching as closely as possible the image of the caterpillar which, in his cocoon, becomes totally liquefied before reemerging with a changed physical structure. If the "essence" can become sufficiently uncovered to share in the forces of evolution and bring the corresponding signals to our awareness, and if the "ego" can become sufficiently fluid to act out whatever role should be demanded in a given evolutionary (dynamic) situation—then man is not only capable of sensing and following the evolution but of becoming its effective instrument, as indeed it has to be accepted as the basic human predicament.

In the following, I shall very briefly comment on some of the approaches I enumerated in Table 3. I have no intention of describing them in any detail, but I wish to sketch a perspective of how they fit the task of "tuning in" to evolution.[1]

The *Way of Knowledge* is the most broadly recognized way of integration. It has also become the major way in Western culture. However, it does not always rise to the evolutionary level. The close linking of the mythological and the evolutionary levels marks times of mysticism. After Western culture has passed through an epoch of mysticism from the twelfth to the sixteenth century, it witnesses now again the rise of a new mysticism characterized by Scholem (1955) as one transcending the framework of a particular religion; it "strives to piece together the fragments broken by the

1. The systematization presented by Naranjo (1972) has contributed helpful ideas here, without relieving me from the task of designing a more explicit and broader system for the purposes of this chapter. Metzner (1971) presents good introductions to some major approaches: astrology, alchemy, actualism, Tarot, Tantra, and the *I Ching*.

religious cataclysm, to bring back the old unity which religion has destroyed, but on a new plane, where the world of mythology and that of revelation meet in the soul of man."[1]

The great revelations about the dynamically evolving order of the universe, the cosmologies and astrological systems, stand for an approach which strives to learn through the understanding of precepts and objective order, usually from the perspective of a heavenly order. Such cosmologies sometimes find their expression in human measure in mandalas (graphic/symbolic representations) and mantrams (accoustic/symbolic representations), especially in India and Tibet. Such mandalas and mantrams are capable of conveying directly a holistic image of cosmic order, of "speaking" to the human body in the same way in which great art does. Among the cosmologies, the *I Ching*, (see: *I Ching*, 1967), often quoted in this book, occupies a unique position insofar as it incorporates the interplay of the orders of heaven, man, and the earth in the most comprehensive and dynamic way.

The opposite approach is Tao in the sense of no-knowledge (not to be confused with tao, the right path). Tao is the total emptying of the mind which links man to his own origins and opens him to the ultimate in nonreflective knowledge as evoked by the chorus in Sophocles's "Oedipus at Colonus":

Not to be born surpasses thought and speech.
The second best is to have seen the light
And then to go back quickly whence we came.

It is an attitude, which has been proposed as a third approach besides rational and intuitive knowledge, not just by mystics but also by reputable scientists (Siu, 1957: Sibatani, 1973). But the knowledge gained in such a way, is not readily communicable. It reaches only a receiving self which resonates with a Taoist attitude.

Astrology has deteriorated to a very superficial idea in our time. Originally, it dealt with energy configurations in the universe which manifest themselves in particular constellations of matter and generate effects beyond those of gravitation and not yet observed by science. Modern physics has confirmed the underlying view that

---

1. One may point here to the appearance of an increasing number of mystic and psychic revelations giving converging perceptions of elaborate evolutionary cosmologies. The most astonishing example is perhaps the very detailed, psychically transmitted writings of Alice Bailey (in particular: Bailey, 1934).

matter is but a manifestation, a condensation, of energy. But astrology goes further in linking evolutionary microprocesses—the dynamics of life and consciousness, which are also manifestations of energy transformations—to these configurations of energy. Thus, astrology links the basic, and empirically evident, phenomenon of biorhythms, as well as the less evident phenomenon of fate, to objective energy constellations which may be interpreted as "cosmic mudras."[1] The configurations of the earth-moon system are clearly reflected in the female sexual cycle in humans as well as animals. The sun-earth system with its one-year periodicity gives rise to the seasons and is linked to many biological and psychological cycles. The more complex periodicities of the whole solar system are the principal subject of classical astrology; modern science has only recently taken some interest in the possibility of correlating the movements in the solar system with complex biorhythms. But a galactic outlook, taking into account the 26,000 year cycle in which the direction of the earth's rotational axis changes, would bring into play the energy distribution in the Milky Way system, as seen from the solar system, and permits the linking of various levels of periodicity. Such an outlook is implicit in the mystic interpretation adopted by Jung (1961) and widely shared among young people, that the present epoch is marked by the transition from the Age of Pisces (the two fish, linked together, also symbolize dualism)—the two millennia under the sign of Christ—to the imminent Age of Aquarius.

Mythologies, like all cults, may be understood as a linking backward, a reactivation of the genetic potential (T. and D. McKenna, 1972). The history of religious beliefs unfolds like a beautifully painted fan whose harmonious design radiates outward from the same point of origin, as A. Huxley's (1950) comparative anthology of writings from the great religions demonstrates so impressively. The same holds for all forms of expression of great cosmological paradigms whether they are formulated in terms of religious systems, mystical visions, or scientific syntheses.[2] They all come into human

1. Mudras are called in Hinduism those "objective" positions of the human body which appear to enhance the communication with the forces of evolution, the sense of movement and direction (see also the ontological approach in Table 3).

2. LeShan (1969), comparing statements made by great scientists and mystics, interprets their congruence as a manifestation of the law of correspondence—the conclusions from the search within match the conclusions from the search without.

consciousness by a process of revelation which seems to activate human monads so as to reflect aspects of the same basic world view. A basic evolutionary orientation seems to underlie most of these manifestations of an intuitive grasp of order and movement. In fact, I believe it would be most rewarding to compare the basic patterns which emerge from the great variety of mystic and psychic messages and writings, often received through media which appear never to have aspired to such roles or studied the complex relations which, in their intuitive systems, all seem to fall into place. To the extent that I have acquainted myself with such literature, I would expect a similarly "unified" result as came out of Huxley's juxtaposition of religious revelations.

The study of the history of religions, as well as of other branches of mankind's political and spiritual history, is still the most valid approach to gaining a sense of what, beyond all cultural boundaries, may be called the dimensions of humanness. I feel that this is particularly true for a study of the arts throughout the ages and across all cultures.[1] It is in the arts that man's building of an inner and outer anthropomorphic world, a home filled with human warmth amidst a cold and objective universe, has found its most comprehensive and profound expression.

A particularly interesting form of knowledge is provided by the Jungian collective unconscious, built up through all the past generations, and its symbolic elements, the archetypes which surface in dreams, fairy tales, and all our symbolic relations to the world. The empirically proven phenomenon of Jungian unconscious implies a learning process of whole mankind, a specific human contribution to the evolution—to noogenesis in particular—which may be an important one. It ties in with the concepts of Mind-at-Large, and of reincarnation, and it points to a principle of conservation of

---

1. I recall one of the most formative experiences of my life which came to me during my first trip to Japan, visiting en route centers of all the great living religions, from Christian and Jewish, Muslim and Hindu, to the Buddhist faith. And there in Japan, in the magnificent Horyuchi temple nestled in the hills surrounding Nara, I suddenly had the feeling I had come full circle, as if something had snapped into place inside me. There, on the other side of the planet, in the wooden statues nearly a thousand years old, I found the same expression of human dignity and kindness, the same depth of humanness, the same knowledge about the meaning of life which had struck me deeply in ancient Greek and Roman statues, in European art of all great epochs. I felt I had finally measured out some essential dimensions of humanness. Ever since, my transcultural meanderings seem to relate to a frame of reference which can hold whatever I keep filling in.

spiritual energy. This does not come as a surprise once there is a unified concept of all forms of energy; we know already that physical energy is conserved.

A recent view (Hillman, 1972) suggests the abolishment of the older notion that archetypes establish themselves in our lives like structural elements and emphasizes instead that it is through their interplay that consciousness takes shape (again appearing as something like a standing wave-form); behavior would then evolve from processes of interaction between archetypes. In such a view, there would be no place for the claim of conventional behavioral science that behavioral patterns remain basically unchanged. Tarot cards, becoming a popular source of wisdom again, may be seen as embodying in the twenty-two Great Arcanes a medieval version of a system of archetypes, with the fifty-six Lesser Arcanes mapping out the field of potential human response—a remarkable expression of a nondualistic view of the determinism versus free will issue. Games with Tarot cards constituted an ancient approach to trace the tao of man in terms of archetypal interplay. Quite generally, fortune-telling appears to have been the prime purpose of all ancient card games.

The *Way of Compassion/Love* manifests itself primarily in approaches to healing. Neurotic behavior has come to be regarded no more as a symptom of malfunctioning (in particular sexual malfunctioning, as Freudians used to claim), but as a symptom of blocked creativity which can be helped to find its proper channel to be acted out. I believe seriously that neurotics and their creative potential, frustrated in the current cultural setting, constitute one of the greatest hopes for an emerging new culture. At least, there is a reservoir of creative energy banging away at the doors which the established culture has closed on change and innovation!

It is significant that the Jungian notion of psychotherapy as a dialogue, as interaction between two persons, is again coming to the foreground more recently, as it befits a technique located at the mythological level of the I-Thou relationship. The evolution of psychotherapy may be followed in terms of the Hindu chakras, the centers for human energy exchange which the different schools of psychotherapy focus on. If Freud was concerned exclusively with the sex (the second) chakra, and Adler focused on the power (the third) chakra (Ram Dass, 1971), we may say that Jungian approaches partly explore man's relations at the heart (the fourth) chakra. Logotherapy, the restoration of meaning in a mythological relationship with God (Frankl, 1949) would then aspire, in a partial way, to the fifth chakra, and spiritual psychosynthesis, the search for the transpersonal "higher self," the "essence" which man shares with the divine principle and which is embedded in the Superconscious (Assagioli, 1971), has perhaps the sixth chakra, that of wisdom, foremost in mind. These latter approaches, dealing with man's relations at the evolutionary level, are but in their infancy at present.

Gestalt therapy, a favorite shortcut to psychotherapy, focuses on the acceptance of responsibility for the manifestations of the whole self, not just the "ego" pattern, e.g., by acting out characters which appear in dreams. The general technique of psychosynthesis, working mainly through imagined situations, is related to this. Personal relationships are at the center of the ever-popular encounter group techniques of which quite a variety is practiced today, breaking up encrustations around the self and giving an experience of communication, if only a superficial one.

The approaches I listed at the evolutionary level, go beyond a personal interaction in establishing a direct contact between a person, a healer, and forces which seem to be drawn directly from cosmic energy constellations and are not yet explainable by current science. This, according to Metzner (1971), holds for actualism, a revival of an ancient art called *Agni Yoga*, and is meant to enable the student "to experience the evolutionary processes and purposes of man" by means of a kind of psychotherapy of the Actual Self, or "essence." It certainly holds for psychic healing, which may be recognized in a wide spectrum of forms, from mere "energy passing" and energizing/balancing massage all the way to the healing or removal of sick organs on the spot, or even over considerable distance. Many people are potential healers of a lower grade. What is acutely felt during any healing process is a kind of resonance phenomenon in both persons involved, linking them to some external (cosmic) "vibrator." It seems that electron emission from the body, probably from acupuncture points, plays a role in these processes (Krippner and Rubin, eds., 1973). A similar phenomenon, the "rising of the Kundalini energy" along the spine toward the head (through the seven chakras in the body), is supposed to result from the practice of *Kundalini Yoga* which focuses on breathing exercises.

Particular interest focuses again today on the vast body of knowledge put into perspective by the ancient system of Chinese internal medicine. It is based on the recognition and explicit interpretation of the law of correspondence between the dynamic orders of the universe and of the human body. Acupuncture is the practical application of this philosophy. A variety of techniques—from needling to moxybustion (the burning of small cones on the skin) and electro-stimulation to Japanese Shiatsu (pressure point) massage—are all based on the same system of acupunture points and meridians (connecting lines of energy flow, or resonance). In Europe, Paracelsus attempted in the first half of the sixteenth century to establish a link between alchemy and medicine and to develop a medicine based on the law of correspondence between the "big world" (the planets) and the "small world" (terrestrial metals); his metal therapy is forgotten today.

Psychosomatic therapy the contemporary Western version of a holistic approach to body/soul/mind, does not go so far as to include cosmic references but also recognizes the subtle interplay of physical, emotional, and psychological factors in the whole person. In some respects, it is in the process of linking up with some aspects of acupuncture. Polarity and zone therapy, as well as reflexology and iridiology (the representation of the whole body in the feet, or in the iris of the eyes), develop particular aspects of psychosomatic therapy.

At the "awareness" level, we find also a whole host of approaches which form the backbone of most of the activity going on in "growth centers." There is a variety of approaches to "grounding" or developing the Kath, the center of instinctual wisdom, in particular with reference to sexual functioning. Bioenergetics is an approach to a therapy of the whole person which focuses on breaking up distortions in the body and muscular structure which usually constitute the other side of the coin of rigidification or distortion of the self—with classical psychotherapy focusing on the corresponding patterns in the mind. Deep massage, structural integration (restoring the original muscle patterns by breaking up the fascia which envelop them and distort, in particular shorten, them over longer periods of life), and structural patterning (optimal balance in standing and walking) are all based on the bioenergetic principle which holds that the whole person and its problems are reflected in the structure and movement of the body. The Japanese approach of developing the lower belly, or center of gravity—called Hara—is related to this (Dürckheim, 1956).

The *Way of Meditation and Mind Control* is most directly geared to the goal of enlightenment. Besides all types of formal meditation techniques, from mere awareness of one's own self to spiritual enlightenment in prolonged sitting in za-zen (non-doing), we find here many approaches which are geared to bringing about the experience of a new reality (see, for example, Naranjo and Ornstein, 1971). At the awareness level, they include the "emptying" techniques of asceticism, sensory deprivation, fasting, and isolation in deserts or underwater tanks (Lilly, 1972). Also, certain biofeedback techniques belong here which are currently being experimented with (especially at the Menninger Foundation at Topeka, Kansas), in particular techniques to put oneself into the calm alpha-wave state (low frequencies of brain currents). On the other hand, there are the "filling" techniques of sensory awareness and the so-called mentations (the placing of one's consciousness as well as of a specific idea into specific parts of the body, e.g., goals into feet and hands, or form into the eyes, and physically living out the consciousness dispersed in such a way). (Lilly, 1972).

At the mythological level, we find certain Tantric techniques, especially "energy exchange" through eye contact and—not a technique, but the "classic" way in which a new reality is gained— schizophrenia, the breaking up of one specific reality and the gaining of a new one through linking backward to the origins of self. But above all, we find centered here the expressions of art, in particular those forms of art which have been cultivated in the Western world. Poetry, painting, music—all transform reality and elevate man's perception to the evolutionary level, at the same time integrating the three levels to such an extent as perhaps no other approach is capable of. As I am writing this, my eyes wander—as they often do—to the simple, exquisite painting of a Cyphoma sea shell hanging near my working desk.[1] At the mythological level it becomes alive as a woman's vagina, breathing, pulsating, opening up. At the evolutionary level it becomes a symbol of life, a guiding image for the owner of the painting in her present phase of personal evolution, an expression of her self-attunement.

Other approaches to enlightenment use drugs which unquestionably have opened up a new consciousness in the younger generation. Drugs also appear to further the integration of all three levels in a particularly strong way. The drug experience need not be continued to maintain this new consciousness, this feeling for a new social and spiritual reality which has led to a spiritual renaissance in America and partly also in Europe. Hallucinatory drugs also serve as an introduction to shamanism, as the recent books by Castaneda

---

1. I have chosen this painting for the jacket design of this book.

(1969, 1971, and 1972) and the McKennas (1972) bear fascinating testimony. A shaman is traditionally called somebody who is capable of moving on the four-dimensional hyperplane to other space/time places.

At the evolutionary level, we find the avowed goal of most meditative techniques, the higher satoris, or higher states of consciousness which are really states of enlightenment in which, at the highest two levels, the four-dimensional reality of evolution may be experienced directly. Satoris also come in quantum jumps (for a description of the higher satoris, see Lilly, 1972). Such satoris are usually experienced spontaneously and may then change people's lives profoundly. Children seem to have them more often than grown-ups (perhaps because of their weaker "ego" crusts), but they are also experienced by many people at the moment of their death or near-death (as surviving people report who have nearly drowned or fallen to death). The Albigenses, a Christian sect living in Southern France, around Albi, in the twelfth and thirteenth centuries, considered heretics and exterminated by the church, seem to have known the secret of inducing the highest satori in dying people. Indian yogis probably can reach satoris at will. The promise of contemporary Western esoteric schools to teach a training which would make it possible to slip in and out of higher satoris at will has not yet been fulfilled and has been changed to a promise of raising the average low every-day satori level permanently. Very sensitive people have been transported by music, in the highest performance quality, to higher satoris—but the truly enlightened spirit of re-creating music also has almost left our world in an age of increasingly commercialized music life.

Esalen massage can put people into higher satoris. This new massage technique, named after the Esalen Institute at Big Sur, California, is practiced as a high meditation and emphasizes the balancing and integration of body, soul, and mind along the lines of psychic healing. It does not so much work on the physical aspect of the body as it does on its psychic aspect, on the "aura," putting the latter in contact with harmonizing forces beyond the two persons involved.

Finally, I should like to include here the ancient Japanese Noh play which is finding larger audiences again among the younger people in Japan. Its essence is an indescribable, unique harmony through a system of chanting, rhetoric, drama, dance, mime, masks, and the extraordinary beauty of the old silken costumes—a harmony which seems to unify the taos of heaven, man, and earth and draws the sensitive spectator into experiences of mystic enchantment.

The *Way of Movement/Action*, the last of the four ways of integration, focuses on the intrinsic values of pure action, without any utilitarian thought given to the fruits of that action. Polanyi (1966) recognized the "tacit knowledge" to be gained from it. At the awareness level, we find such approaches as chanting, drumming, gymnastics, Hatha Yoga (the physical exercises of yoga), movement and movement therapy. Dancing, especially folk dancing, is an old approach here. One may also recall the European approaches to eurhythmics earlier in this century, and the rhythmic movement games which are experimented with in Germany in the context of learning and healing processes for children (for example, the "Schulspiel" of the composer Carl Orff). In America, one finds more recently among younger people much folk dancing, as well as the joyous Sufi dancing (which mostly uses dances created by a Sufi group at San Francisco in the ancient spirit). We find here also the Japanese martial arts, in particular archery which is practiced as a meditation, as well as calligraphy, a popular meditation in the Far East even in our hectic present.

The dancing of the dervishes in Konak, which arises from prayer and flows back into prayer, would belong to a higher, probably evolutionary level. Integration of the way of action through dancing —really through "meditation in action"—seems to have also played an important function in the Greek Dionysian mysteries. The exquisitely balanced sequence of movements in the ancient Chinese Tai-chi chuan which may be seen practiced on the lawns of the Berkeley campus every evening, tries to express movements belonging to higher orders. The same expression of the law of correspondence may be found in the ancient Japanese tea ceremony (not often performed any more in its pure and unabridged form).

At the mythological level there are many versions of role playing techniques, theater games and psychodramas. They purport to enhance spontaneous expression in action. Primal therapy, probably much overvalued, aims at a kind of liberation through the "primal scream."

At the evolutionary level we find alchemy, together with technology the expression of man's irresistible urge to change the world. As Eliade (1962) points out, the technological itch may be interpreted as a degraded form of man's age-old dream of changing the world through psychic energy, of becoming the master of superhuman forces available in the spiritual uiverse. Alchemy was the great hope based on the spiritual correspondence between the microcosmos of man and the macrocosmos of the universe; technology is the contemporary reality of man's naive belief in his domination over nature by applying physical force. Alchemy aimed at the regulation of evolutionary forces; technology is content with replacement.

We also find here rituals and certain forms of drama. Greek tragedy was a means for "tuning in" to evolution by combining the (objective) Apollonian demand to recognize oneself in the conditions of human existence, and the (subjective) abandonment to Dionysian ecstasis in a total transformation of the self. The spirit of ecstasis (literally, "standing outside oneself," that is, shifting to another reality) has been at the core of the Dionysian mysteries, and animated, and still animates, the rites of many cultures. In the West, it is precariously alive today in some of the experimental theaters, in particular in Jerzy Grotowski's Theatre Laboratory in Wroclaw (Poland), in Eugenio Barba's Odin Teatret in Holstebro (Denmark), and in the Bread and Puppet Theatre which has recently found a home in Plainfield, Vermont. These theater groups are really schools, communes, spiritual communities, and work groups all in one, which have found a creative vehicle for their personal and group growth. Other esoteric schools/communes focus on training outsiders more or less comprehensively in their approaches. A conspicuous example in America is the Arica Institute with headquarters in New York and teaching houses in many other cities, operating since 1971 and presenting a rich package of ancient Sufi and Far Eastern techniques redesigned "for the needs of contemporary Western man" (see Lilly, 1972). The older Arcane school, with centers in New York, London, and Geneva emphasizes the formation of "spiritual groups" which never meet physically. Quite generally, as I have discussed at the beginning of this chapter, group experiences may be expected to enhance the capability of "tuning in" to evolutionary movement which, in turn, becomes an aspect of collective creativity.

In the following seven points Naranjo (1972) summarizes the general outcome of taking any way of integration:

—Shift in identity (putting the "ego" in its proper place in the functioning of the total self);

—Increased contact with reality;
—Simultaneous increase in both participation and detachment;
—Simultaneous increase in freedom and the ability to surrender;
—Unification—intrapersonal, interpersonal, between body and mind, subject and object, man and god;
—Increased self-acceptance; and
—Increase in consciousness.

The sense of direction, which we have set out to look for, streams primarily from the increased contact with reality, unification of subject and object as well as man and god, and increase in consciousness. Even if this sense of direction comes to us as a general experience of movement only, not readily translatable into vectors of action, it will nevertheless guide our consciousness and the processes by which it projects ideas onto the appreciated world and exercise a gentle, but powerful influence—as powerful as the rigid inherited templates which shield our consciousness from the ever changing processes of evolution. The clearer our sense of direction and movement, the more meaningful can design become, but also the more powerful. Because only when we sense all these powers around us, all this vast stream of evolution, can we coordinate and use them for our purposes—which will then also be the purposes of overall evolution.

The same dangers that lurk in physical technology, also seem to befall some of the applications of "human technology." These are the dangers which are inherent in a lack of measure. Measure is needed in the balance between "natural growth" and "gardening," between the horizontal learning process called life and the added vertical learning processes that make use of technology. The American faith in specialization and technology, in handing over problems and buying solutions, runs the risk of leading into the same fragmentation and "problem-solving" spirit as physical and social technology. The same hidden traps which are waiting for the builders of an outer world, also wait for the builders of an inner world in which man can feel at home. These traps are called sequential (non-systemic) problem-solving and linearity. Falling into them would imply grave distortions of the inner world, comparable to the damage now becoming visible in an outer world built in a technological, not a social, mood. The naive belief that the established rhythm of ancient training techniques can be accelerated, so far results in forms of training that seem to foster egoism, irresponsibility, and superficiality rather than love and qualities that might work, hopefully, toward the unity of mankind.

On the other side, the learning processes outlined above would be most essential for the renewal of the institutions of society. Schooling is still primarily geared to the narrow areas of rational knowledge alone and to a partial exploration of the rational level; it helps less and less to cope with the reality of an integrated framework of life. Education ought to develop along ways of integration as I have outlined them. In this context, experimental programs of high schools in conjunction with the Esalen "growth center" ought to be mentioned. In Chapter 13, I shall come back to questions of what I shall then call inter- and transexperiential education.

At this stage, I am not able to delineate the complete sadhanas, or ways of integration, for the present situation and the corresponding predicament of mankind. However, some of the following chapters will deal with partial integration, mainly linking the rational and the mythological levels. Chapter 11 on models and myths will tie in to some extent with the Way of Meditation. Chapter 12 on a systems approach to planning will deal with a partial integration along the Way of Action and with getting away from a fixation on grossly utilitarian targets. Chapter 15 on aspects of social policy will continue this. Chapter 13 on transexperiential integration will elaborate along the lines of the Way of Knowledge. Chapter 14 on policy design, and particularly the Epilogue on planning and love, will evoke aspects of the Way of Compassion/Love.

In the last two chapters, I explored the ways in which consciousness evolves through perception, taking an "outer" and an "inner" way. In the next chapter I shall discuss the balancing of consciousness, involving both reality and the appreciated world. In this balance, the focus shifts from perception to apperception.

(10)

CENTERING CONSCIOUSNESS

The following aspects of human experience emerge as essential inputs to our consciousness:

—what we *are* in our biological, genetic, anthropological, psychological, social, and cultural nature—composed of inborn as well as conditioned aspects;

—what we *feel* in a sensual as well as an affective way, ranging from physical pleasure and pain all the way to compassion, love, hate, joy, sadness, and that all-embracing feeling of communication with life, *eros*; we feel in the interpersonal domain as well as through resonance and dissonance with our natural, social, and spiritual environment, through communication with nature as well as with the man-made world;

—what we *perceive* through sensation, intellectual and intuitive cognition and conceptualization, but also through "direct insight" of an instinctive or mystical nature;

—what we *know* through perception and learning, but also in the form of mythical knowledge, through initiation and revelation, and through whatever is left of our instinctive knowledge;

—what we *want* through satisfaction of drives and desires, needs, motivations, and aspirations, through the realization of values, but also through self-expression in a general way and in a wide variety of forms;

—what we *conceive* through our will, on the basis of our understanding and under constraint by our biological, psychological, social, and cultural nature which comes into play with its basic as well as its conditioned elements;

—what we *can do* in terms of our basic capabilities, skills, modes of expression and communication—semiotic modes, but also more generally iconographic modes.

All these experiences touch our whole self—body, soul, and mind. As we do not perceive only through the senses, we also do not conceive only through our mind. We think with our body, we feel with our mind, we perceive with our soul, too. The above structure is not supposed to prefigure any sectoralization in dealing with human experience. Rather, it is supposed to give a rough idea of a larger system—a *system of total human experience*—in which knowledge and the functions of brain and mind are embedded. Indeed, any idea of a sectoral approach breaks down when we realize the extent to which all these aspects are interrelated: what we are expresses itself in knowledge through a variety of scientific disciplines, such as biology, psychology, or anthropology, but it expresses itself also in what we feel, perceive, want, can do, and conceive. Similarly, our feelings translate themselves into what we want or conceive, but also through intuitive formulation into what we know. Bringing into play different forms of expression and communication, variations of what we can do, provides completely new perspectives not only to what we feel, know, want, and conceive, but even to what we are as a socially and culturally conditioned being. And so forth.

The first four types of experience—what we are, feel, perceive, and know—dominate in the encounter of consciousness with reality, whereas the rest—what we want, conceive, and can do—emerges more visibly from the encounter with the appreciated world in the model/myth feedback cycle, which I shall discuss in the following chapter. However, this division of emphasis is by no means absolute. What we want is also experienced in the physical, social, and spiritual needs which reality seems to put to us; what we perceive and know is considerably influenced by the adventures of our imagination interacting with the appreciated world, etc.

Consciousness, or that part of self which is active in the human design process, is shaped by feedback processes on two sides, those linking it to reality and those linking it to the appreciated world. What, in Chapter 8, I called the represented context of human existence, is the fruit of *perception* in a general sense, or, in Leibniz's terms, of the inner state of the human monad which represents the outer world. Comparable to a standing wave pattern emerging from the interference of feedback processes impinging from both sides, from reality and the appreciated world, consciousness may be said to arise and take shape from *apperception*, or the reflective understanding of this inner state. All social change is connected with the development of a new view of reality through apperception; recognition of self as a process rather than a solid structure is also breaking through in modern sociology: ". . . if one wants to ask who an individual 'really' is in this kaleidoscope of roles and identities, one can answer only by enumerating the situations in which he is one thing and those in which he is another" (Berger, 1963).

Through apperception, consciousness gains a dynamic dimension of change, a forward thrust in a specific direction. Such change occurs in mutations to new dynamic regimes, which are brought about in the same way that I discussed in earlier chapters as a characteristic of dissipative structures. The pattern of change is the same, with inputs by the processes of apperception generating instabilities which lead to a mutation, to a new dynamic regime at a higher level. Keleman (1974) has described this process as a basic sequence which characterizes all processes of self-formation, especially those of children: excitation and charge build-up (new inputs through interaction with the outer and inner world)—containment of charge (increase in self-perception)—giving up boundaries (new mode of self-expression). Self, or consciousness, seems to exhibit the same dynamic characteristics as other nonequilibrium systems in the physical, biological, and social domain—a striking argument for the hypothesis that "order through fluctuation" (see Chapter 3) is a basic and universal principle of evolution, at work also in the human domain and in the design process.

The richness of the formative processes which shape consciousness and make it mutate, give expression to new contents, is a measure of the openness of self to the processes playing in the world and transcending the mere human realm. This richness may be linked to *charisma*, that holistic and dynamic quality in people which, as Max Weber already recognized, moves the world. Charisma, in this perspective, is a concept that reaches far beyond any power- or knowledge-based concept of elitism. It is not the result of a social constellation or of individual ambitions, but of evolutionary dynamics finding expression in individuals as well as in social interplay.

Modern psychology has made considerable progress in elucidating some of the patterns which constitute our inner nature, in particular through the recognition of an interplay of archetypes. The foundations for a "quantum theory of man's psychic world" have already been laid by C. G. Jung. But still we know next to nothing about perhaps the most important aspect of our biological nature, namely the genetic and psychological potentials and limitations of our adaptability to changes in the environment, changes which we often initiate through our capability for action. In other words, we lack the knowledge for balance, for regulation. We are used to brushing

aside any concern expressed on this matter by asserting that man will
always be able to adapt—on no better grounds than the obvious
fact that he has so far been able to adapt, although his genetic
heritage has not changed significantly over the three hundred generations which have passed since the Stone Age. Biologists estimate
that fifteen percent of man's genetic heritage is "in use" to cope
with a given type of environment, and that the rest constitutes our
reserve for adapting to change. But the New Stone Age of contemporary urban life may have used up much of this reserve already
—the openly neurotic behavior of people in New York or Tokyo
is a danger signal which may no longer be overlooked.

Obviously, our experience of what we feel is in crisis. Sexual
frigidity is but one symptom of it. All our relations with the world,
based on feeling, seem to be passing through a somewhat cooler
zone. The full meaning of *eros* in its wealth and all-embracing
dynamics has almost left Western culture—it may still remain with
a few artists devoted to a "communion of love," to the *Liebesge-
meinschaft* between artist and public of which Furtwängler (1955)
spoke at the end of his life. In Chapter 14 I shall interpret this sign
as a characteristic of a time of transition, of a shift in focus from
social to cultural organization.

When I watch children, mainly up to an age of perhaps five or
six years, I always receive some confirmation for my belief that
man is "naturally" made to perceive the wholeness, the unity in the
physical as well as the spiritual world. Fairy tales, as with all poetry,
express and enhance such a perception of wholeness. But in our
scientific-technical age, children are subsequently schooled in such
a way that they perceive the world increasingly in a fragmented
way; they are prepared to think and express themselves in the
terms of science—with its inherent fragmentation of reality—and
to act in the mode of technological problem-solving, whether the
issue is a technological or a human one. This is but another consequence of suppressing man's deeper roots in nature. "Im Ganzen
liegt das Dämonische" (In the whole lies the demonic), Goethe
knew.

In the realm of what we know, we generally exclude myths (or
we think we do) and common sense. These forms of basic experience were very important in the earlier phases of mankind's psychosocial evolution; they worked well for many old cultures and still
do for those contemporary smaller-scale cultures which we call
"primitive." We know so little about ourselves, and even the little

we know is hardly applied to any design purpose except perhaps to advertisement and other forms of "manipulation of the subconscious."

Ambitions govern what we want to a far larger extent than drives, which Western culture tends to suppress or "elevate." Graceful self-expression has become difficult in our time, which values competition and conformism in behavior so highly—it has even become rare in a physical sense, with ambitions and the will to force things leading people to artificial grimacing and self-conscious movement—to games played in the expression of the body no less than in mental terms.

What we conceive has become pitifully narrowed down in the traps which we have set for ourselves in our linear economic and technological growth world. The systems approach, which I shall outline in Chapter 12, attempts to restore some of our inborn wealth of conception.

Modes of expression and communication, finally, which form the most important aspects of what we can do, have become greatly impoverished in our time, too. We have become almost sterile in exploring and developing basic anthropomorphic modes of expression and communication—by "anthropomorphic" I mean "made to the measure and expressing the basic capabilities of man." Language, symbols and icons, forms of artistic expression, and sexuality are less innovative and alive than they were at the peak of this and other cultures—and I am not prepared to confuse permissiveness with aliveness. Even more trivial modes, such as fashion, cooking, and celebrating are becoming ever less imaginative. We are used to praising the creations of the past and to speculating how long we might still enjoy them. Old paintings, antique furniture, and objects from other cultures are valued much more highly than our own creations. With the exception of France and Hungary, and perhaps also Italy, creation in cooking seems to have come to an end in Europe, not to speak of North America. On the other hand, we have shifted our interest to technomorphic modes of expression and communication, "made to the measure and expressing the capabilities of technology" in the form of computer languages, electronic music, television commercials, and so-called TV dinners.

All these basic experiences can be animated, varied, explored, and brought into play, consciously or not, through the human purposes implied in a living, dynamically developing culture. "Left to themselves," they tend to deteriorate toward the extremes on either side. The challenge thus lies in developing a balanced attitude or, as it may be called in a more comprehensive sense, in centering consciousness.

Consciousness forms the mental and spiritual core of human life; it interacts with all aspects of the life of an individual as well as of a human system. Therefore, it has to be centered in all dimensions in which human life expresses itself. In their most generalized representation, these dimensions of centering are fourfold: the forward direction of life at the three levels of perception and inquiry in which it expresses itself, namely the rational, the mythological and the evolutionary level; and the vertical dimension linking these three levels.

As I pointed out in the last chapter, the same four dimensions were also recognized in ancient Chinese philosophy and were used to build the *I Ching*, the Book of Changes. The interplay of the four pairs of opposites characterizing these dimensions gives rise to a comprehensive spectrum of change encountered in the life of human individuals and systems:

In ancient times the holy sages made the Book of Changes thus:
Their purpose was to follow the order of their nature and of fate. Therefore they determined the tao of heaven and called it the dark and the light. They determined the tao of the earth and called it the yielding and the firm. They determined the tao of man and called it love[1] and rectitude. They combined these three fundamental powers and doubled them; therefore in the Book of Changes a sign is always formed by six lines.

If the law of correspondence—"as above, so below; as below, so above"—may be assumed to hold, then man shares integrally in the ways of heaven and of the earth. His life is an expression of the tao of heaven and of the tao of the earth no less than of the tao of man. A human life is lived at all three levels simultaneously and must therefore be centered in its movement along all three of them, finding also the right balance in their interplay.

It might be asked whether centering should not be discussed in terms of physical, social, and spiritual space rather than in terms of levels of perception and inquiry. Human life, indeed, finds expression also in triads of physical, social, and spiritual aspects. But it would be mistaken to equate the tao of the earth with physical,

---

1. In the sense of "compassion."

the tao of man with social, and the tao of heaven with spiritual aspects. The tao of the earth, for example, is a physical concept at the same time as it is a social concept—as we are learning in the age of understanding our relations with a "whole earth"—and a spiritual concept as the place in which we root, become "grounded." Similarly, the tao of man and the tao of heaven, in relation to human life, have to do with physical, social, and spiritual aspects. A discussion in terms of the three taos, or the corresponding levels of inquiry, allows the expression of the integrity of human life, and of the life of human systems, in a more profound way.

In Figure 14 I attempt to depict the dimensions of centering

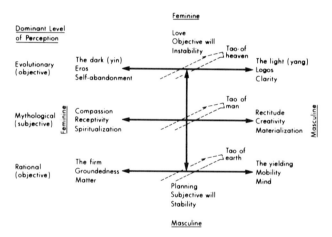

Figure 14. The four dimensions of centering : the forward direction of the three taos, or right ways—heaven, man, and earth—and the vertical dimension linking them. Man lives all three taos simultaneously.

graphically and to identify them with the pairs of opposites mentioned in the *I Ching*, as well as with other key notions. At the evolutionary level we find the archetypal pairs of opposites which govern the myths of Dionysos (self-abandonment) and Apollo (clarity), of Isis (the moon, *eros*) and Osiris (the sun, *logos*), of the inseparable yin and yang. At the mythological level we find the opposite prototypes of human attitude exemplified in this book by Giraudoux's Suzanne (compassion, process) and Robinson Crusoe (rectitude, structure). And at the rational level we find the basic opposites which I have already introduced in Chapter 2, "grounded-ness" (matter) and mobility (the "ego trip" of the mind).

The vertical dimension is the one in which the creative tension between love (the destabilizing, "fluidizing," and upward-drawing force, embodied in man's energy-dissipating "essence") and planning (the stabilizing, "hardening" tendency, embodied in man's energy-focusing "ego") is lived out. I shall go into these aspects more deeply in the Epilogue on "Planning and Love" where I shall also identify pairs of opposites with the feminine and the masculine principle, objective and subjective will, as I have already indicated in Figure 14 for completeness' sake.

I do not intend to explore here in much detail what the notion of centering implies at the three levels of perception. However, in Table 4 I attempted to list key words suggestive of attitudes which emerge from the basic experiences I have enumerated above. For each type of basic experience, and for each level, I tried to grasp the essence of the two extreme attitudes as well as of the centered attitude which does not just balance between the two extremes but embraces them both.

It would be quite wrong, I believe, to identify the side of nature or grounding with what we are, and the side of the mind with what we make of ourselves. There is no reality behind any concept of a "natural man" from which modern man has deviated and to which he ought to be redirected. Nevertheless, Rousseau's notion, in *Emile*, that "nature has made man happy and good, but society deprives him and makes him miserable" has crept up occasionally in psychological and social literature; Marcuse's *Eros and Civilization* (1955) presented perhaps the most profound and interesting discussion of this dualistic view. Man experiences not just as a natural, but also as a social and cultural being, formed and influenced by many aspects of his world—including those aspects he has himself brought into his world, such as technology, growth, information, social organization, myths, and models.

It seems significant that at all three levels in Table 4 the rubric for what we feel contains the ancient Chinese notions by which the opposites were characterized. The ancient approach was much more one of feeling than of analysis. Today we focus much more on what we want, conceive, and can do; in short, on the experiences streaming primarily from the mental process of model and myth formation and less from our encounters with reality in a more direct way.

Centering is perhaps most urgently needed today in the dimension of what we want. It makes a vast difference whether what we want expresses itself as expectations or as hope. As Illich (1970 a) remarks, hope is subject to potential satisfaction, but expectations—at least the familiar "ever rising expectations" which are but "a euphemism for a growing frustration gap, which is the motor of a society built on the coproduction of services and increased demands"—engender growth and fixation on progress instead. The belief in the possibility of rational action itself is anchored, beyond rationality, in hope, whereas both lack of hope and fixation on expectations lead to irrational behavior. One does not need to be an optimist to plan and act, or to fulfill merely what one considers one's duty—but one needs to have hope.

Expectations have become the cancer of both advanced and developing countries. Focusing on quantitative, not qualitative, growth they tend to smother hopeful evolutionary developments, even acting against life. Hope is with life; it gives life a chance. The "serious business of living" which *ikebana*, the Japanese art of flower arrangement, celebrates in a flower, finds its support and strength in hope. In Gorki's *The Lower Depths*, the flower on the wall in the barren surroundings of a suburban slum becomes the symbol of resurrection from numb apathy, from the life of the *Lumpenproletariat* which has even become devoid of suffering, toward tender love, to the hope for life.

In the domain of what we can do, the extremes at the evolutionary level may be characterized by the figures of Don Quixote, the tragic but noble activist who bravely fights for the real good, only to find himself helplessly entangled in unreality; and the man of Tao, the serene inactivist, who does not judge between right and wrong, who totally abandons his own self in order to become "true-self." Both positions, worthy in themselves, do not provide valid models for all mankind in the present phase of evolution, in which human design has become of crucial importance. The centered attitude is one of subtle regulation, of identifying with the evolutionary flow and thereby acquiring the possibility to use the powers available in it and through it.

Centering in the vertical direction means fully assuming the human role between heaven and earth and integrating all three levels of which human life is an expression. It is particularly revealing to look

TABLE 5

Vertical integration of the centered attitudes of Table 4.

LEVEL	WHAT WE . . . ARE	BASIC EXPERIENCE					
		FEEL	PERCEIVE	KNOW	WANT	CONCEIVE	CAN DO
Evolutionary (tao of heaven)	Evolutionary agent	Communion (truth)	Order of Process (Evolution)	Purpose ("know-where-to")	Hope (dynamic security)	Syntony, "tuning-in"	Regulation
Mythological (tao of man)	Cybernetic actor	Communication (morality)	Process	Goals ("know-what")	Expectations	Roles	Action
Rational (tao of the earth)	Self—(Becoming)	Continuation (beauty)	Change	Efficiency ("know-how")	Possession (static security)	Force (replacement)	Behavior

TABLE 4
*The centering of basic attitudes at the three
levels of perception/inquiry. These attitudes
give expression to characteristic sets of values.*

BASIC EXPERIENCE	FEMININE	←	TYPICAL ATTITUDES CENTERED	→	MASCULINE
			EVOLUTIONARY LEVEL		
are	"Empty channel" ←		Evolutionary agent	→	Rebel (Prometheus)
feel	Eros		← Communion (truth) →		Logos
perceive	Attraction		← Order of Process (evolution)		← System
know	Flow, change of regimes	←	Purpose ("know -where to")		→ Defined regime
want	Self-abandonment ←		Hope (dynamic security)		→ Clarity
conceive	Finding		← Syntony, "tuning-in"		→ Invention
can do	Inactivism (Man of Tao)		← Regulation (con- trol of powers)		→ Activism (Don Quixote)
			MYTHOLOGICAL LEVEL		
are	Drifting (Suzanne)	←	Cybernetic actor	→	Homo faber (Robinson Crusoe)
feel	Compassion	←	Communication (morality)	→	Rectitude
perceive					
	Gestalt (quality)	←	Process	→	Structure
know					
	Dynamic forces	←	Goals ("know-what")	→	Connections
want					
conceive	Nonattachment	←	Expectations	→	Demands
	Adaptation	←	Roles	→	Investment
can do	Receptivity	←	Action	→	Creativity

RATIONAL LEVEL

are	Creative (Being) ←	Self (Becoming)	→ Ego (Doing)
feel	Instinct (groundedness, the firm)	← Continuation (beauty)	→ Intellect (mobility, the yielding)
perceive	Contents	← change	→ Form (measure)
know	Viability	← Efficiency ("know-how")	→ Utility
want	Basic drives, needs	← Possession (static security)	→ Achievement of targets
conceive	Conservation	← Force (replacement)	→ Progress
can do	Instinctual response	← Behavior (learned response)	→ Leverage (technology)

at the vertical integration of the centered attitudes as they apply to the three levels, respectively. Table 5 provides an overview over what may be called some of the centered dimensions of humanness.

One of the most powerful motivations in human life, *fear*, assumes quite different expressions at the three levels. At the rational level, we fear being delivered to impersonal, anonymous system pressures. This kind of fear dominates, for example, the "Club of Rome" publications on world dynamics (Meadows, et al., 1972). Also, individual life appears as unique at this level; therefore, the death of an individual appears as the extinction of a unique entity of life and consciousness. There is no god at this level. At the mythological level, fear is primarily related to the feeling that a precarious homeostasis with the environment is breaking down due to changes in the environment which we can neither foresee, nor master, nor adapt to; the image is that of a capricious god who does not listen to our appeals. But here, fear relates more to the possible extinction of mankind than of individual life. At the evolutionary level, finally, there can only be the fear of being "left outside" the main flow. The fear of death disappears here in both the individual and the mankind perspective and yields to trust in the overall process of noogenesis and evolution.

Table 4 is not just an aid for gaining a panorama of actual and potential attitudes. It constitutes nothing less than *a system of values*, which are embodied in the enumerated attitudes. I have already argued, in Chapter 8, that values are lived out in the tension between opposites and that they form a complex system of such pairs of opposites, not all of which are in play at a given time. Part of the structure of such a system becomes visible in Table 4. Pairs of opposite values hang together in the same way in which the seven forms of basic experience hang together and penetrate each other. If the three parts of Table 4 are stacked on top of each other, a hierarchical order emerges. It is the same hierarchical order which has been recognized by man in many systems of thought: the true (sometimes split into the sacred and the true, which appears redundant) at the evolutionary, objective level; the good at the mythological, subjective level; and the beautiful at the rational, objective level. The useful which is sometimes added to this list, hardly represents a separate hierarchical level; it stands for a slant toward the masculine side. The same holds for the rational, sometimes mentioned as a primary value.

But just as human life is lived at all three levels simultaneously, there is no absolute order of values in terms of a rigid hierarchical pyramid, with the top values governing the lower echelons. As in human systems, hierarchical order here has to be viewed as a multi-echelon system, geared to the coordination of expressions at all levels.

A full human life makes use of a value system which is "activated" to a high degree, bringing into play as many pairs of opposites as possible and centering them horizontally and vertically. Each key notion listed in Table 4 stands for a whole host of values pertaining to different aspects of the same basic attitude. Emphasis on "Homo faber," for example, brings subvalues such as entrepreneurship or work *per se* into play and brings other values of the system, such as rectitude, expectations, competition, or rationality to resonance. Human life is essentially the activation of values at all levels, a notion also becoming recognized by modern psychology (Burkhardt, 1973). The fuller the life, the bigger the challenge of centering. An unbalanced life tries to pursue values which may be located at different sides from the center; it tends to bring both opposites of other pairs to resonance which creates irreconcilable situations and unresolvable tension. A centered life is characterized by a highly activated and harmoniously resonating value system.

In Chapter 8 I said that values are the norms evolving from the encounter of self/consciousness with reality, specifically at the mythological level, and in the social realm. In this chapter, the notion of value has acquired much broader meaning as the effective core of attitudes that reach across all three levels and that reflect man's relations to heaven and earth, and to himself.[1] In the end, the value system constitutes a profound expression of man's reflective or nonreflective response to all aspects of what, ultimately, turns out to be his very own life, the tao of heaven—or evolution, as we may say now—realizing itself through man no less than the tao of the earth, or organic life. In Chapter 13 I shall come back to this value system and outline how it enters into the design of a human world.

Consciousness becomes centered through interaction with reality and the appreciated world. It contributes to building the appreciated world by forming models and projecting them onto an emerging imaginary world which thereby becomes "appreciated." It is time to look at some of the approaches which allow the spirit of centering to enter these processes in building the appreciated world. In the following chapter, I shall discuss the process by which consciousness invents, builds, substantiates, and changes the appreciated world, a malleable construction of mind and feeling in which we live out our hopes and expectations, our beliefs and faiths, our aspirations and desires, our objective and subjective needs, and our personal style before we bring all this to bear on the hard world of reality. This is the feedback loop that links the formation of models and myths and that has the power either to stimulate or stifle human action in significant ways.

1. Relations, in turn, constitute a much broader concept than choice, which is often used to define values in a utilitarian perspective. Compare, for example, Kluckhohn's (1951) frequently quoted definition of values as "a conception, explicit or implicit, distinctive of an individual or characteristic of a group, which influences the selection from available modes, means, and ends of action."

PART IV

The Appreciated World Evolving

I leave you to the pleasures of your higher vices,
Which will have to be paid for at higher prices.
T. S. Eliot

THE appreciated world is the product of our subjective self. It consists primarily of models projected onto it by consciousness—models of the world as it is and as it ought to be, and which return and govern consciousness in the disguise of powerful myths. To keep the model/myth cycle always open and flexible, to perceive and deal with reality through a fluid spectrum of models, is the most difficult task at the core of creative design. Two approaches may help in this task: a systems approach to planning, aiming at centering the appreciated world in various dimensions; and the idea of inter- and transexperiential inquiry, organizing man's total experience (the totality of formative factors of consciousness) toward a purpose. Both approaches focus primarily on the elevation of inquiry from the merely rational level of quantity to the mythological level of quality. The latter approach may also be viewed as a basis for a future fuller and more human education.

(11)

OF MODELS AND MYTHS

The feedback link through which man with his consciousness relates to his appreciated world is perhaps the most crucial one. It is here that the alternative between positive action and mere self-contained behavior is prepared. It is here that the whole rational framework for the justification of such action is welded together. But it is also here that the rational, the broader experiential and the plain irrational come together in greatest intimacy, touch and interpenetrate each other. It is the feedback loop of models and myths.

I use the term model for any static or dynamic concept which is supposed to represent, predict, prescribe, or simulate structures and processes of reality; and the term myth for any model which becomes effective as a basis for further design and action. In this book, I am primarily concerned with dynamic models, whether they represent fixed or changing structural relationships. Fixed structural relationships are characteristic of growth models restricted to mere quantitative growth. But design and generally all cybernetic types of interaction imply structural change. Models of structures tend to quench processes which are to be kept alive, in the model as well as in reality. Therefore, it is better to model processes first and let them define tentative, changing structures. This modeling of processes is the general field of *forecasting*.

Model building is essentially a creative or inductive activity by which we are building a generalized system out of elements of our consciousness—though not just of observations and rational knowledge, as science would have it. Myths become effective through an appreciative or deductive process by which we impose such a generalized system on our consciousness where it structures our attitudes and valuations, the whole appreciation that—again through a model —we bring to the world. Models and myths are inextricably interwoven; what is conceived as a model in one direction becomes a myth in the other, and governs as such the building of new models.

So closely interdependent are the two parts of the feedback process that we may call them a short-circuit within man's mind. If this sounds alarming, it *is* alarming. Of all the traps built into man's cybernetic existence, the model/myth loop is the most difficult vicious circle to break out of. Locked into this trap, we resemble the young man of whom Heinrich von Kleist tells in his essay "Über das Marionettentheater" (On the Puppet Theater): When he had posed for a sculptor, the graceful young man felt so flattered by his portrait in stone that forthwith he tried to control his movements in a studied attempt to match the grace he had discovered in his replica. He started to contemplate his image in a mirror for hours and days. But it was as if an invisible net strangled the free play of his gestures. His movement lost all its previous gracefulness and, within a year, no trace was left of the charm and the inner life which previously had been a source of joy for all people who came in contact with him. Similarly, self-reflection on our creative model building, through the mirror of myths, may impair the gracefulness of our creative life.

Model building is an activity going on continuously not only within individuals but also in social and cultural systems. The activities of governments and corporations include a great deal of model building concerning the internal working of the organization as well as its interaction with the physical, social, and spiritual aspects of its surroundings. All information-processing social systems generate dynamic models, not all of which may become explicit, though. So also do cultural systems. Special skills are needed to build an explicit model as well as a mental, social, or cultural organization. Humans and human systems have evolved specifically in such a way as to function as builders of models and myths.

But the information processed is not just made up of knowledge. It includes the elements of consciousness which evolve primarily from the interaction between consciousness and reality—what we are, feel, perceive, and know—as well as those elements which evolve primarily from the interaction between consciousness and the appreciated world, in the model and myth forming processes themselves. These latter elements of experience include what we *want*, what we *conceive* (the model-forming capability itself), and what we *can do*. They are introduced, tested, weighed, and modified through the appreciated world by means of our ideas of what the world can do for (or to) us and what we can do for (or to) the world.

Both models and myths are man-made stilts with the help of which

we may temporarily lift ourselves out of the stream of evolution—
our minds primarily, although we tend to follow with our whole
self—and orient ourselves as to which direction to take. In those
moments in which we *are* the flow, we need neither models nor
myths. But most of the time, and in accordance with the specific
human condition, we need them for two important reasons. First,
we cannot bring our full human potential, our capability for design,
into play if we abandon ourselves fully to the stream. In order to
become effective, to play the active role evolution demands of man,
we have to use our capabilities of reflection, judgment, and invention.
Action is linked to some form—a temporary, changing form under-
going continuous metamorphosis, but nevertheless recognizable to
work on. Vagueness leaves room only for passive endurance and
behavior. In our mythological relations, we have to gain at least
an idea of where we are and where the stream is carrying us if we
want to try to steer a course in it, to act. We are compelled to make
an effort toward predicting, within some range, the course of evolution—although we know only too well that evolution cannot be
predicted, only lived. This results in one of the basic dilemmas inherent in the human predicament, a dilemma which again may be
interpreted in terms of the energizing tension between the yin and
yang, the listening and the acting. We cannot abandon ourselves
passively to evolution if we want to take our role in it seriously;
but with every step of emancipation we risk working against evolution. Evidence is abundant in human history that models and myths
were, indeed, often at variance with tao, the right evolutionary path.

The second reason for which we need models and myths is simply
that every human action takes time. Since we can meaningfully act
only in the framework of a model and against the background of a
myth, we have to assume explicit models and myths in order to be
able to design and carry out our actions. We shall never fully reach
for ourselves the ideal of regulating a totally fluid world, of "walking
without touching the ground," as the Yoga ideal of full enlightenment is described. In the present stage of evolution we still touch
the ground and are even supposed to; we walk on stilts, not too
gracefully, and we leave deep traces in the ground. This is why we
have a need for assuming a structure for the ground and a myth
which tells us that the ground is firm enough to carry us.

All models and myths are man-made, creations of the human
design process. This statement holds for models and myths in the
physical as well as in the social and spiritual domains. It will be felt

by many as shocking, again. So desperate is our dependence on viable models and myths that we subconsciously try to elevate them from the muddy world of human emotions and interests to the crisp, clear heights of absolute truth. Physical models are set absolute, and the myth of science is created. Social models are set absolute and their corresponding myths, or ideologies, subsequently imposed by social habits and taboos, or also by intimidation and force. Spiritual or cultural models are set absolute and their imperialism is defended on the grounds of belief in the form of religion or other evolutionary or pseudoevolutionary myths. Language, also a spiritual model, rigidifies and locks us even more firmly into a cultural myth than men can do.

And yet all these models and myths are human artifacts, carrying the imprint of our physical, social, and spiritual consciousness which, in turn, has developed in close contact with our spatial and temporal environment in all three domains. We keep forgetting this. Had we not forgotten, all the passions, conflicts, and triumphs of human history, political as well as cultural, would not have come into play. Much of the energy active in the lives of humans and human systems is polarized and flows between "electrodes," represented by opposites in models and myths. But if the tension between opposites energizes humans and human systems, models and myths may also be viewed as focusing devices for this evolutionary force. Sutherland (1973) points to the significant circumstance that the paradigms of the human sciences do not have the expected unifying and simplifying effect, but that "the existence of one seems to spawn the existence of an essential inverse."

A human attitude toward models and myths is itself played out in the tension between two extremes. The poetical image of Giraudoux's Suzanne stands for one extreme (see also Chapter 5). She is totally responsive to the life stream, virtually dissolves in it. Therefore, she is wary of fixing her own position through models and myths. Robinson Crusoe's brave and dull figure stands for the other extreme. Utter mobility versus utter rigidity, living out versus building, compassion versus rectitude, process versus structure. Neither of the perfectionist extremes constitutes a viable general proposition for life in mankind's current situation, although Western man is playing Robinson Crusoe all over the world. The "generalized Heisenberg principle," to which I already referred, is at work here again: either total movement, or total fixation—or in between a balance of imperfect movement and imperfect fixation. In this light, imperfection appears as the very essence of human life. But to center ourselves between the two extremes means that first we have to go a long way to link hands with Suzanne. It means that we need the courage to be imperfect.

Dynamic systems modeling is inherently based on assumptions which are rarely made explicit. The most important of them pertain to the following questions:

—What is the nature of the system to be modeled? A basic distinction may be made here by using the level of consciousness of the system in question and the ways in which it relates to its environment. In more familiar terms, this leads to the basic distinction between physical, bio-logical/social, and spiritual (specifically human) systems.

—What is the position, in relation to the system, from which the model is being constructed? Is the system being viewed from the outside, from the point of view a member of the system might take, or is it viewed as a whole and in the context of its total environment (as a member of a metasystem)? These views correspond to the rational, mythological, and evolutionary levels of inquiry, as outlined in Chapter 5. The dynamic aspects of the system appear as change, or process, or evolution (unfolding wholeness), respectively.

—What are the parameters in which the outcome is to be obtained? Since they determine the parameters of the whole model, and thus of the inputs to be used, a normative idea is introduced with their choice. What is considered to be maximized or minimized, or kept stationary or at some equilibrium, implies already a certain notion of what is considered as "good" and what as "bad." The higher the level of consciousness, the more important this is, and particularly important with human systems.

In the following, I should like to propose a simple characterization of dynamic systems models by using the basic distinctions mentioned under the first two points above, and arranging them in a 3 x 3 matrix (Table 6). A look at the nine fields of the matrix, corresponding to nine different approaches to dynamic modeling, permits a clearer understanding of the potentials and limitations of known approaches as well as of yet unsatisfied aspects of the full challenge implied by the ambition of modeling human systems. Each one of these nine model categories pertains in some way to the human world, but no single model belonging to any of these categories is capable of ade-quately representing parts of the human world in which human faculties are played out to the full extent. Only by applying different approaches *in the proper perspective* may we hope to comprehend

TABLE 6

A matrix of dynamic modeling approaches. Progressing from the lower left (field S_1) to the upper right (field A_3), a hierarchy of dynamic systems principles emerges, all of which play a role in the human world.

Perspective (level of inquiry) / Nature of System	Rational Change	Mythological Process	Evolutionary Evolution
Apperception Self-reflective consciousness (ternary)	A_1 Rational role-playing models	A_2 Learning approaches, gaming, moral systems approach	A_3 Formation of spiritual paradigms, cultural morphogenesis, syntony
Perception Reflective consciousness (binary)	P_1 Mechanistic system models, rigid behavioristic models	P_2 Adaptive system models, homeostasis, ecological models, contingency models	P_3 Inventive system models, biological and social morphogenesis, co-ordinative models of relations
Subjectivity Nonreflective consciousness (singular)	S_1 Newtonian mechanics, cause/effect models	S_2 Irreversible thermodynamics, statistical mechanics, econometrics	S_3 Dissipative structures

Diagonal axis labels (bottom to top): Causality (local determinism) — Probability (statistically pre-programmed behavior) — Vitality (life-preserving behavior); Volition (moral action); Creativity

Left-side axis labels:
← Physical systems →
← Biological/social systems →
← Spiritual/human systems →

in a better way micro- and macroaspects of the integral human world.

The *nature of a system* may be characterized by its capability of relating to its total environment (which itself may be composed of physical as well as social and spiritual aspects). Three basic ways, or levels, may be distinguished here, following concepts developed in Chapters 6 and 7 (compare, in particular, Figure 11):

—A nonreflective, or "blindly unfolding" type of consciousness which may be called *subjectivity*. All natural systems, from atoms to galaxies, organisms and societies, have subjectivity (Laszlo, 1972b). The activity of *physical systems* may be fully described at this level.
—A reflective type of consciousness which interacts with reality through *perception* in a binary scheme, absorbing experience and forming a "represented context of reality." Communication is of the signal/response type. The specific activity of *biological systems* (which are inherently *social systems* at some level) may be described in terms of relations formed in such binary systems of consciousness and reality.
—A self-reflective type of consciousness which reflects also on its own role in the game through *apperception*. It becomes possible by the introduction of a "memory" in the form of an "appreciated world." Such a ternary scheme (see Figure 11) forms the basis for an auto-catalytic unfolding of consciousness, or *creativity*. *Spiritual*, and in particular *human systems* may be more fully described only in terms of apperception.

The *perspective*, or *level of inquiry*, represented by a dynamic model, may be characterized by the already familiar categories of the subject/object relationship:

—In a *rational* view, interference with the dynamics of the system is effected from the outside by imposing *control*.
—In a *mythological* view, *regulation* from inside the system is effected largely by negative feedback action in relation to the environment, generally by attempting some sort of homeostasis. Competition, or Darwinian (external) factors of evolution are emphasized here.
—In an *evolutionary* view, the focus is on the unfolding of wholeness through a morphogenetic chain, utilizing energy in manifold ways through positive feedback. Accordingly, the evolutionary view empha-sizes internal or coordinative factors of evolution.

Depending on what aspect of the human world is supposed to be modeled, approaches pertaining to all nine fields of the matrix may be usefully applied, provided they are viewed in the proper perspec-tive. The boundary conditions for their application become even clearer, when we proceed across Table 6 from the lower left (field S_1) to the upper right (field A_3), spanning a variety of *dynamic systems principles* from causality to creativity (finality). We find five levels of such principles:

1. Field S₁: *Causality*, or *local determinism*, which is often effective in describing microaspects of change in a nonsystemic and short-range perspective. Newtonian mechanics belongs here as well as many predictive cause/effect models in the social domain.
2. Fields S₂/P₁: *Probability*, or *statistically preprogrammed behavior*, through which may be studied the dynamics of systems for which no potential of self-renewal is assumed ("linear" forecasting). In physics, field S₂ accommodates irreversible thermodynamics and statistical mechanics, whereas in the social domain the bulk of econometric models, based on statistical correlations of past behavior (and implicitly perpetuating such behavior) belongs here. Most of the models applied so far assume equilibrium conditions and define probability accordingly. The recognition of nonequilibrium conditions in natural systems, for example in nonequilibrium thermodynamics, leads to vastly different concepts and to the morphogenetic principle represented in field S₃. The approaches corresponding to field P₁ may be termed *mechanistic system models* since no change in system structure is allowed for; in fact, the system is assumed to act in analogy to a Skinner-box, determining the behavior of its inmates (e.g., no food increase—no population increase). Many types of behavioristic models belong here, from input/output models to most of the published versions of "System Dynamics," including its application to world dynamics (Forrester, 1971; Meadows et al., 1972).
3. Fields S₃/P₂/A₁: *Vitality*, or *life-preserving behavior*, adaptive to changes in the environment, which has so far been at the focus of General System Theory and has served well to illuminate adaptive, and particularly homeostatic, behavior in the biological and to some extent also in the human social domain. Field S₃ refers to dissipative structures which secure their potential for entropy production, or life in a general sense, by mutating to ever new dynamic regimes ("order through fluctuation"). What is modeled here—not in a predictive way, since the fluctuations leading to mutations may themselves be random—is evolutionary, morphogenetic dynamics in the physical realm.

Field P₂ accommodates approaches to model *adaptive systems*. Ecological models of all kinds, for the nonhuman as well as the human world, belong here, as well as various types of *contingency models* in the social realm. Whereas most models so far take an equilibrium-oriented approach, more realistic nonequilibrium approaches focus on boundaries (Holling, 1973), and on the conditions for metastability (Prigogine, 1974b). Contingencies in social models may be introduced either through man-technique interaction, comparing simulation runs of mechanistic system models which use various sets of initial assumptions or constraints (e.g., in the world dynamics studies by Forrester, 1971, and Meadows et al., 1972), or by building a spectrum of contingency responses into the model itself—as has been done for various applications of "system dynamics"—or by combining the single-level "system dynamics" approach with a multiechelon approach, "ruling" the responses from a superior coordinative level (Mesarović and Pestel, 1972). In any case, the approach is restricted to a behavioristic outlook, limited by the imagination of the model designer and/or operator.

Field A1 is characterized by rational models of conscious man—the *homo economicus*, *homo politicus*, and other clichés explicitly or implicitly underlying much of current social theory. Here belong the simplifying models of game theory, but also more sophisticated approaches which assume fixed human roles (goals) and adaptive controllable variables in the system, allowing for internal and external fluctuations.[1]

4. Fields P3/A2: *Volition*, which may be considered as the underlying principle of both *moral human action* and (on a much slower time-scale) of morphogenesis in biological evolution. It expresses freedom in the focusing of available energy.

Field P3 accommodates models of biological morphogenesis (e.g., the topological approach by Thom, 1972) as well as *inventive system models* in the social realm (see Chapter 4). No social models of this type have yet been proposed, but the idea would be to formulate something like a principle of "social order through fluctuation" and the conditions for the morphogenetic evolution of systems of relations. Generally, this type of model might be called a *coordinative model*, but I would not expect that all dimensions of human relations can be expressed through one model.

A multimodel approach is also essential for field A2 where all kinds of group-learning approaches belong, especially gaming (which, strictly speaking, is not a model, but a "live" simulation experiment). Generally, this is the field for a *moral, or humanistic, systems approach* which attempts to embrace as many views as appear relevant to a systemic situation (see Chapter 12). An equilibrium approach would seek consensus (often impossible to reach), whereas a nonequilibrium approach would emphasize the recognition of conflict as most essential (Cleveland, 1972).

5. Field A3: *Creativity*, or the mutation of human consciousness in morphogenetic ways, generating (or "retrieving" from an unconscious potential) ideas and forms, plans and models, values and motives, works of art and science, physical, social, noetic, ethical, and religious systems concepts. It is responsible for man being not only a stabilizer of his world, but also a "crisis-provoker." In the domain of science, ideas for such a mutating "order of knowledge through fluctuation" have been proposed by Kuhn (1962) and, more recently, by Laszlo (1973a). What is badly missing so far is a useful concept for the mutation of values, or, in more general terms, for the formation of new cultural paradigms and complex systems of such paradigms (religions, ideologies, myths, world views, etc.)—in other words, a model of the

---

1. Work on such "adaptive market-adjustment models" is in progress 1973/74 at the Second Institute of Physical Chemistry (Prof. Prigogine) at the Free University of Brussels (Marshall Moffat, oral communication).

"Zeitgeist"—as well as a concept for the model of communication which leads to a wide sharing of these paradigms and beliefs ("syntony"). An understanding of the internal (coordinative) factors in the evolution of human consciousness will probably become possible only in the framework of a wider theory of evolution which recognizes consciousness as a manifestation of an overall unfolding of energy.

These five dynamic systems principles may be understood as forming themselves a hierarchical stratified order, ascending from causality to creativity, from a microscopic mechanism to a macrosopic unfolding touching the numinous; ascending enriches meaning, descending detail (if at the price of reductionism). A tight, predictive model may be constructed only for a world reduced to causality. The farther we climb the hierarchical ladder, the less is one model capable of grasping the fuller dynamics of human systems. Vitality can already only be partially represented through a model to the extent that the model designer or operator is capable of anticipating the spectrum of responses which may come into play. Volition and creativity become even more elusive to modeling and cannot be represented by single models any more. However, they may be approached to some degree through the study of a multitude of models representing causality, probability, and vitality aspects in the light of different assumptions, temperaments, descriptions, anticipations, feelings, and insights. Thus, the effort of modeling the human world becomes itself the *design of a dynamic system of approaches* extending over all fields of the matrix in Table 6. The more fully we grasp the human world in such a way, the less it is prediction we seek, but an understanding of processes and evolving order in the human world, a recognition of human capabilities as well as a "tuning-in" to evolving order embracing, but going beyond, the human world.

Most models applied so far to the human world do not go beyond a representation of probability and are geared to the ideal of equilibrium and stabilization. Yet, in a time of transition, such as ours, volition and creativity govern the human world ever more importantly. The task of modeling turns into one of helping the anticipation and realization of social and cultural morphogenesis. The dynamics of the energy unfolding in these realms may be grasped only by recognizing the nonequilibrium nature of human systems and internal coordinative factors in the evolution of relations and human consciousness. There is a wide-open field for empirical research as well as theory-building.

Behavior is predictable, action is not. The current behavioral fashion in model-building reveals one of the profound reasons for the sore state of science in general. Whereas action is a moral activity, behavior is an amoral one. Who simply behaves, cannot be held responsible. Nor can an amoral, i.e., value-free, science be blamed for any of the outcomes of its application. A statistical determinism has to allow for random deviations. But what we do not accept for our social code of right and wrong, we find no difficulty in taking seriously under scientific disguise. That the message of man's existence "beyond freedom and dignity" is taken seriously and is even jubilantly received in some quarters today, is almost beyond my belief. That scientists should take pride in reducing human beings to mere response mechanisms in Skinner-box type environments, is perhaps the ultimate caricature of man's relations to his models and myths. We hear, and not for the first time, of scientists whose interest lies in proving their models, not in understanding reality. Myrdal (1958) has mercilessly exposed social scientists of this type. Econometrics is another and even more powerful manifestation of this unreal attitude of scientists running amuck. We can make the human world predictable only at the expense of derobing it of its specific humanness. Our successes in technology have seduced us into applying technological and mechanistic concepts and "problem-solving" procedures to the human world.

We touch here perhaps one of the most powerful myths active in the Western world, a myth which belongs to the spiritual sphere. This is the myth that man is not capable of bearing his freedom, that it has to be interpreted for him, that he needs a model imposed from the outside to cling to, that he is not free in his own creative capability of building models and myths. In Chapter 1 I quoted from Dostoyevsky's tale of the Grand Inquisitor, in his *Brothers Karamazov*, which expresses this myth most profoundly.

The memory of the Grand Inquisitor came back to me most vividly when I spoke in Moscow about forecasting and long-range planning. Within certain constraints we are free to shape our own future, I told my audience of predominantly young people, who sat on the edge of their chairs, electrified by these prospects. As usual in Eastern countries, no hand went up in the official question and answer period, but afterwards I was surrounded by many of my listeners who were in a state of great excitement. One of them asked me what was on all their minds: "I can shape my own future! But who is going to tell me what to do?"

I believe that, to a certain extent, we need such protective myths —at least as long as we do not, in a global way, feel the forces of evolution in sufficient strength to place our confidence in them. Religions used to shield us from too much freedom, from too heavy a moral burden. In the shadow of this shield grew amoral science and "neutral" technology. But modern social science is discovering that "freedom is not available rationally" (Berger, 1963), that it is excluded *a priori* from the causally closed system as the human world appears at the rational level.

It is perhaps significant that the young rebels who bravely fight some of the myths suffocating Western life, become themselves an easy prey of narrow and dogmatic myths, be they relics of a model conceived in the past century (such as Marxism), or vague idealizations of something happening at a great distance (Maoism). The moral urge of their fight is only too often readily abandoned for the lure of security in amoral and even outright deterministic myths.

The feedback interaction of amoral models and myths is bound to lead to excesses in those evolutionary phases that are characterized by quantitative expansion instead of qualitative change and which Teilhard de Chardin (see Chapter 3) linked to end states of developments, announcing major evolutionary mutations. We are living in such a time. The interlocking of models based on ever-rising expectations, with myths guaranteeing an automechanism of self-regulation comparable to Adam Smith's "hidden hand," creates a positive feedback loop in which amoral aspirations and beliefs form a self-exciting system. The recent worldwide discussion around the report of the "Club of Rome" on the limits to growth (Meadows, et al., 1972) is a case in point. On the one side there was the suggestion of a myth which holds that the dynamics of the world may be modeled in one single model at the rational level of inquiry, based mainly on physical processes which occasionally reflect behavioral assumptions but cannot grasp any role-playing or spiritual processes —and that a simple change in the initial assumptions would adequately represent the processes through which massive policy change might come to be conceived and implemented in the real world. On the other side there was the indignant outcry of social scientists over the lack of confidence in the market mechanism. Had the latter not been *defined* so as to take the moral burden from us, to automatically ensure self-regulation without human, that is to say moral, intervention whatever we do? And the cool arguments of others who simply held that the quantitative assumptions of the underlying model were poor, that the precise location of the limits and therefore also the time when they would be reached were misjudged and that consequently we need not discuss any of this until we had approached the limits more closely—which, at an average doubling time of ten years of many of the factors involved, would not mean more than the possible "gain" of a decade or two of inaction, of continued amoral processes of growth.[1]

Böhler (1972) has recently drawn attention to a mechanism of "hardening" models into myths which is at work in the formation of expectations. It has its origin in man's mythical need for security, which itself goes back to fear. The search for security leads to the assumption of regularities (myths) which permit the formation of forecasts by logical deduction. But logical deduction is always linear and partial by its very nature, since it tends to isolate a single aspect of reality. Thus, "all forecasts are hypothetical partial models, whereas aspirations and reality are total in character, embracing an infinity of single aspects." A discrepancy always remains between them and implies that fear before the obstacles on the path into the future would not be removed. This, in turn, increases man's urge to reduce insecurity by turning the rational institutions of culture, namely technology, economy, state, and mass media into boundless mechanistic growth processes. The Utopian urge for material security thus becomes rationalized as "economic reason," where, as Böhler states "it is the most outrageous unreason."

Models of a dynamic world deal with processes and orders of processes. A mere structural model is of no relevance if it does not, at the same time, specify how structure is meant to generate or regulate processes.

Morgenstern (1972 a, b) has recently pointed out that economic theory is still primarily concerned with physical objects, such as output, stocks, and returns, and does not deal with that kind of activity that *moves* "the economy," namely decisions. An economic theory focusing on processes, according to Morgenstern, would

---

1. For a representative response by conventional social science, see the opinions of the Science Policy Unit at the University of Sussex (Freeman, et al., eds., 1973). A much more imaginative response, taking an evolutionary systems perspective, may be found in: Laszlo, ed. (1973 b).

introduce a reflective relation between theory and object of theory—
in other words, and in the terminology of this book, would be based
at the mythological level of inquiry.[1] Economy, remarks Böhler
(1972), is in its essence not about material aspects, but about status,
recognition of opinion and person, power and strength, self-expression, and ultimately about the magic inherent in the man-made
world. To understand economy, we must understand man's spiritual
world.

There are at least some approaches which help us to break out of
the vicious circle of interlocking and short-circuited models and
myths. The development of long-range planning has not only extended the time range of planning, but shaken up its entire framework and methodology. In the next chapter, I shall outline this new
spirit of planning in the framework of a systems approach. Forecasting, in particular, has found a new perspective and methodology
(Jantsch, 1972 b) which may be put to good use for systematically
"energizing" and "exciting" the model/myth cycle to develop new
qualitative alternatives. In this respect, a breakthrough has been
achieved, at least in theory, which seems to tie in well with the
general task evolution has given man in the current situation.

With this development of forecasting away from the mechanistic
and behavioral models of conventional social science, and toward a
fuller use of the human design capability, the nature of the processes
underlying the formation of models and myths is changing significantly. If we may still call them inductive and deductive, then
only with the understanding that these terms have now acquired a
much broader notion than is implied by a conventional scientific
method. In the inductive process of model building we still go from
the particular to the general—but no longer basing merely on empirically available observations as science does, nor bringing into
play only the elements of human consciousness, but all processes
swinging in the overall design process. In other words, the formation
of models is no longer the exclusive affair of consciousness projecting
itself onto an appreciated world in a linear and causal process (or,
with myths, in a single feedback loop), but an integral aspect of a
much subtler and much more complex process of a whole system
balancing itself out—consciousness, appreciated world, and reality,

1. Morgenstern sees the basic distinction between natural and social science
in their location at different levels of inquiry, the rational and the mythological,
respectively. I see the three-level structure of inquiry as cutting across all
domains, physical and social as well as spiritual.

all in their threefold physical, social, and spiritual aspects. And the deductive process of relating to a myth may be seen in an analogous way—no longer working merely by logic, which is the deductive method of rational science, but also by feeling and by insight which are the "deductive methods" applicable to mythological and evolutionary inquiry.

The mythical nature of planning reveals itself in the widely shared expectation that planning will remove uncertainty and complexity. If planning is viewed as an essential aspect of the creative design process, it can only have the opposite effect—uncertainty increases because new options are generated along with honest chances for their realization, and complexity increases because planning widens our view of the infinite complexity of reality. In an evolutionary perspective, the jump toward higher complexity in our conscious view of the world, and particularly in our appreciated world, means the mutation of our appreciative system toward new dynamic regimes, toward ever renewed spiritual life.

The mythical need for security cannot be fulfilled by making a clockwork out of a reality filled with the forces of life and evolution, and least of all the social and spiritual aspects of reality. We need, above all, ways to emancipate our processes of model/myth formation from the rational level of inquiry, to elevate them to the mythological level where quality and total experience comes into focus. The systems approach and the concepts of inter- and transdisciplinarity, which may be expanded to inter- and transexperiential modes of inquiry, constitute approaches in this direction. I shall discuss them in the following two chapters.

(12)

A SYSTEMS APPROACH
TO PLANNING

The appreciated world is the domain of planning, the area in which the design process molds and remolds the worlds that can be and the worlds that ought to be. It is the domain in which we put things together, design thought systems and try to make them work. In the technological age we apply mostly technological modes of thinking and design to conceive and build systems in the appreciated world. After all, they have been very successful in technological tasks. In the early sixties, the state of California assigned four tasks of complex sociotechnological systems design to the technological systems-builders of the aerospace industry to demonstrate the exchangeability of social and technological systems tasks; the attempt failed in many respects, of course.

Of course? Why of course? Because, as I have pointed out in Chapter 5, we live in a world of qualities which emerge from personal feedback relationships with our environment, with other people, with objects, houses, trees, and birds. But our technological systems know-how refers to the rational level of impartial observations, to a mechanical construct in which partial problems can be solved and shelved away, in which pieces hold still when they are put together. Linear and sequential know-how from the rational level does not work at the mythological level of feedback interactions; mere quantity cannot model quality.

If we want precision and logical perfection, we remain trapped at the rational level. But if we allow for the imperfection of bona fide approaches to an elusive reality, not all is yet lost. We can try. In this chapter and the next, I shall discuss two general approaches, or methodologies for striving from the rational toward the mythological level. Since the mythological level is the level on which most of our social life is settled, these approaches come into the foreground of interest at a time in which new approaches to social policy become imperative.

Churchman (1968 b), who has developed the basic philosophy of the systems approach as I shall discuss it here, has concluded "that *the* systems approach really consists of a continuing debate between various attitudes of mind with respect to society." It may be described as a generalized dialectic process for the generation, communication, and processing of verbalized or quantified information. It can accommodate both rational knowledge and feeling, but in being restricted to the information-processing aspects of a system, it is capable of approaching only certain facets of the life of a human system; in particular, it excludes the views to be gained at the evolutionary level—it has to assume a sense of direction which is given.

The central notion in a systems approach is that only the whole system of society—or even of mankind—together with its environment can provide us with the elements, points of view and opinions, knowledge, and, above all, the purposes which will enable us to attempt the design and building of a viable society. A traditional view of planning and management focuses on the internal events in small- or medium-sized social systems with seemingly well-defined boundaries. These boundaries are rarely questioned in a profound way. After all, organizations such as business firms, government agencies, universities, trade unions, and women's associations usually have a clearly recognizable membership, or in most cases even payrolls and organizational schemes to identify it. Thus, the interface between the system and its environment looks like a pretty sharp line.

Of course, the environment is not forgotten in management literature and symposia. Organizations, according to a widely quoted definition, are social systems with specific purposes. These purposes, or at least part of them, have to do with some form of interaction with the environment—business firms are least likely to ever forget that. But some doubt begins to creep in when we read that organizations are supposed to have "specific" purposes. In other words, they allegedly know perfectly well *what* they want, and management has the primary task of determining *how* to do it well—the famous "know-how."

But how do organizations know just what it is they want to achieve? Students of business and management schools as well as astute readers of the professional literature will have the answer ready: Organizations know what they want because that knowledge may be assumed to be *given*. It is implicit or even explicitly spelled out in rigid institutional credos, constitutions, charters, and more broadly in a total cultural background which derives from a specific cultural tradition which we assume to hold for the whole world when we speak of business, government, and education. We may put it this way: The institutions of business, government, education, as well as other institutions play roles which are assumed as given and unquestionable. Presumably, the role of business, and particularly industry, lies in optimizing mass production of goods and services, that is, in optimizing the production side of the economic system, which is the system dealing with the producer/consumer relationships in society. It is supposedly not the role of business to interfere with the consumption side—if business offers a product or a service, society (whatever this notion stands for here) can always take it or leave it.

In other words, business deals with a subsystem of the economic system, which in turn may be viewed as a subsystem of a larger societal system. In doing so, it applies the criteria and measures holding for its proper subsystem, namely, chiefly economic criteria and measures of production—cheapness in price and often in quality, too, high profit, economies of scale, automation, growth in sales and profits, competition, increase of market size and receptivity, swamping of the world with the goods and services business thinks up for society. As one drives to Berkeley on the San Francisco/ Oakland Bay Bridge, the neon sign on top of a paint factory comes into view and catches the eye of the traveler. A big pot of red paint pours its contents over the globe, until the paint runs down thickly on all sides and drips off from the South Pole; the show is accompanied by the triumphant statement: "Sherwin-Williams paints . . . cover the earth!" Here is an outstanding example of the same ugly image which most of management literature attempts to draw of the purposes and roles of all business, and to which management education usually gears its stereotypes.

But how could a whole system's view be put to work instead, in the concrete terms of a business corporation's activities—and for its planning for such activities? After all, business covers only a sector of societal activities, and a business corporation is restricted in scope and often also geographically. How can it go outside its "natural" environment, the marketplace, and what sense would this make? And would it really make that much of a difference at all? The answer is not self-evident to someone not used to thinking in terms of systems—someone who is, will be prepared for vast differences, indeed.

Suppose we take a look at the basic operations of any industrial corporation, systems-oriented or not. We might try to depict them in a rotary scheme such as in Figure 4. However, for the purposes of this chapter, a hierarchical representation as in Figure 15 seems

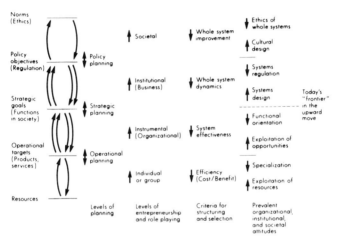

Figure 15. The vertical centering of activity and purpose through the systems approach. Moving to a higher level of discourse permits ordering of activities and concepts toward a purpose; the purpose itself is subsequently questioned from a still higher level.

to be of advantage, since it emphasizes organization toward a purpose. What would really fit best, is a spiral winding itself from the rational to the mythological level; but its representation would be too complex and perhaps confusing.

Elaborating on the hierarchical scheme of Figure 15, we may then say, at the lowest (nonautomatized) management level the corporation organizes resources in such a way as to attain certain quite well-defined *operational targets*, typically technical products. There is a variety of possibilities to choose from in selecting and combining the proper resources, be they raw materials, funds, people with their skills and ambitions, scientific and technological findings, instruments and tools, or physical facilities. What makes sense for selection is mainly determined by the target aimed at. A vast amount of intellectual activity has gone into exploring and structuring ways to organize inputs efficiently, which here normally means economically. Most of what conventionally is called "planning" belongs here, be it network techniques, optimization through operations research and decision theory, risk analysis, economic analysis and discounted cash flow approach, or predictive techniques and models. The operational target, assumed fixed, provides the point of perspective, toward which the inputs and their processing are oriented. We may also say the operational target is at a higher level of discourse and thus permits the structuring of the elements at the lower resources level in a purposeful way—as always, when one succeeds in finding a higher level of discourse and/or abstraction.

We may then envisage the thought process leading to the realization of operational targets, such as industrial products, as sketched in Figure 15 in the lowest feedback loop. The upward reaching arm of it brings into play the availability of a wide spectrum of resources, the downward reaching arm introduces the teleological orientation toward what is supposed to be achieved with these resources. It may be seen at once that only an iterative or feedback thought process makes real sense; that is, will succeed in focusing on a target as well as bringing into play the resources and processes best suited for its realization. Whereas the upward reaching arm focuses on the *exploitation* of available opportunities, the downward reaching arm *limits*, *selects*, and *structures* them. If there is a perfect match between a product idea and available resources, the feedback loop is complete in one instance. This complementarity of approaches is characteristic for all the feedback loops of Figure 15 which I shall go through in succession—it is the essence of the full-scale *normative planning process*, as the ensemble of thought processes sketched in Figure 15 are also called. We may even go further and observe that rarely will the feedback loops be run through in one direction only; rather, the matching process may be depicted as continuous oscillation in both directions, as running through tiny feedback loops within feedback loops within feedback loops—until a match is reached. In other words, the thought processes in normative planning are of the same self-realizing, self-balancing nature I discussed in Chapter 2; they are typical evolutionary processes.

Nevertheless, one or the other of the two directions may dominate in some sense, the upward direction of exploitation of resources, or the downward direction of product ideas. If the upward direction dominates, we find a prevalent attitude of *exploitation of resources*, as exemplified by the history of petroleum companies: from lubrication and illumination through various energy applications, the raw material petroleum became exploited in the framework of one and the same company up to its utilization for petrochemistry, pharmaceuticals, textiles, and recently even food (Single Cell Protein). Even more frequent is the exploitation of personal ambitions and skills, in which case those targets are selected for realization with which people identify most strongly and most aptly; *individual entrepreneurship* expresses itself in this way within large organizations. If the downward direction dominates, we find the prevalent attitude of *specialization*, chiefly in the form of making use of specialized skills and product "know-how" which have been brought together or developed in a firm. Vertical product lines are examples of such specialization in larger organizations.

But to stipulate that an operational target, for example an industrial product, is feasible on the basis of available resources, and that its realization is actively sought, is not a sufficient rationale for the selection of such a target. So far we have dealt with the input/output system of the corporation only. To know whether it makes sense to put a product on the market, we must look at the company's marketplace—the existing as well as the potential one. In other words, we have to look at the environment with which the organization is accustomed to interact. This is the subject of inquiry for the next higher feedback loop in Figure 15, linking operational targets to *strategic goals*. A goal is no longer a well-defined point which can be attained; rather, as we find it in various sports, it is an area where many people can score in different ways—in the upper right-hand corner as well as in the lower left-hand corner and in the center—and repeatedly so. A goal does not go away; it cannot be attained once and forever; it marks an area of focal interest to which many contributions are possible at various times. Typically, such a goal may be represented as a function to be performed in society, for example transportation, communication, power generation/distribution/utilization, food production/distribution, health care, and so forth. The significance of going again to a higher level of discourse is easy to comprehend: we may now structure the selection of industrial products by the function they serve in society. Typically, a function may be served by a variety of technologies. The difference in attitude which started to split thinking in one of the big car manufacturers demonstrates the gain in perspective: whereas half the managers there held the view, "We manufacture cars, as we have always manufactured cars," the other half started saying "We move people"—which brings into play at one big sweep the whole of transportation technology to be assessed and to choose from. But to choose from by what criteria? So far we can say: by feasibility criteria, which are brought into play "from below," from the lowest feedback loop—and by functional criteria brought in by the next higher feedback loop. If I say "functional," I mean here not the internal functions within an organization—manufacturing, marketing, sales, accounting, etc.—but the external functions, those which are served in society by the activities of the organization; as already said above, in the higher loop we go deliberately outside the organization. Functional criteria then deal with how well a certain product serves a function, as compared to alternative products using perhaps quite different technologies, and what its introduction does to a system of human life—for example, what the car technology does to the system of urban life as compared to subways, monorails, bicycles, moving sidewalks or other forms and combinations of urban transportation technology. This already implies going beyond the immediate environment of the marketplace, because the implications for non-buyers matter here as well as those of buyers. Maybe they matter even more.

Whereas the lowest feedback loop may employ crude cost/benefit criteria in assessing economic efficiency, the higher loop calls for the much more sophisticated but currently also much vaguer system-effectiveness concept. What product is most effective in performing a function in a system of human life, while at the same time least disturbing to that system? The instrumentarium of formal planning is already much weaker at this higher level. A social system effectiveness concept is almost totally lacking. But the development and application, in recent years, of technological forecasting techniques has focused mainly at this level and has furthered the systematic search for alternatives as well as their purposeful structuring.

If, in this higher feedback loop, the second from the bottom now, the upward reaching arm is more prominent, we have a type of industrial firm which dominates today in the United States, and partly also in Europe; it focuses on the *exploitation of opportunities*, as exemplified by the widely diversified consumer product companies. The typical American selling slogan "Try something new and different!" is one of its manifestations, and "aggressive marketing" an uglier one. If we can build a supersonic transport, let's do it. As a "shake it and see" approach, the decision to go ahead with the SST has been defended by those countries which are still engaged in its development; they will shake and we will see, presumably. But there are also other serious implications of this onesided *instrumental entrepreneurship*, as we may call it, even if it has become obvious to most people that there may be something profoundly wrong with the naive and boisterous claim "What's good for General Motors, is good for America." A more subtle manifestation is the "rule of the technostructure" (Galbraith, 1967) —individual entrepreneurship, fascinated by a purely technological challenge (such as fancy and complicated weaponry), pushing its way up to determine the behavior of the whole firm and forcing itself onto society or the customer acting on behalf of society, namely the government. In Galbraith's model it is not the top executive nor the shareholder who determines organizational behavior, but the mid-level manager and the key technologist.

If, however, the downward reaching arm of the feedback loop is made more prominent, we find one of the most interesting recent developments in industrial organization, the *function-oriented corporation*. It develops its goals in functional terms and it seeks to integrate as many activities as possible which tie in with a particular function, as well as to develop unbroken chains of activities leading to the service of that function. It even structures its internal organization according to functions and subfunctions. This attitude emerged together with the concept of corporate long-range planning and may be observed mostly in the United States so far; it opens a broad view into the future which used to be obstructed by narrow product-line thinking. It also ties in with the still somewhat fuzzy notion of "social responsibility" of the firm, as it became more fashionable in recent years, since it brings flexibly into play a wide spectrum of technological options for comparative assessment according to criteria of system effectiveness. Some branches of industry have always been more or less function-oriented; for example the power and telecommunication industries, disguised under the technical-disciplinary name of electrotechnical industry. The power industry, in particular, groups a vast variety of technologies under its function, and even such unit outputs as turbogenerators are composed of technically quite heterogeneous elements such as

TABLE 7

Forecasting tasks in a systems approach, as structured in terms of vertical centering. All of these tasks have to be carried out with horizontal (system-wide), time/causality (exploratory/normative), and action/morality (forecasting/ planning/decisionmaking/action) centering in mind.

PLANNING LEVEL	FEED-BACK LOOP	FORECASTING TASKS	HYPOTHETICAL EXAMPLE
Norma- tive	Upper	Changes in values and norms, and in the overall appreciative system (cultural basis)	Global sharing of respon- sibility for world development, improved distribution (closing "gaps"), etc.
	Upper/ Lower	Anticipations (normative future system states); dynamic system behavior for alternative policies	Maintaining sufficient food production for world population; inter- action of food production and population with possible aggravation of world instability.
	Lower	Institutional role playing and invention of new roles	Institution of business orients itself toward concerns for social sys- tems instead of product line development; in- dustry as "planner of society" in interaction with government and higher education (universities).
Strategic	Upper	Implications for policies of dynamically evolving goal patterns	Extension of food produc- tion by nonagricultural technologies shifts global system pressure on population growth to other functions (health, etc.).
	Upper/ Lower	Goals (functions in society); systemic implica- tions and "systems effectiveness" of strategic options (e.g. effects of	Food production and distribution; nonagri- cultural food production technology holds promise of easily added capacity, no burden on land use, little ecological threat,

TABLE 7 (*continued*)

		alternative technologies in social systems)	etc., but requires innovative approach to distribution (new service roles for industry?).
	Lower	Instrumental (organizational) role playing; spectrum of strategic options (e.g. technological "decision-agenda")	Petroleum and food industry link up to develop integral approach to single cell protein on petroleum basis; others develop fresh water algae technology, etc; university research shows new ways, etc.
Operational	Upper	Implications for strategic goals (functions) of dynamically evolving patterns of targets	Impact of single cell protein on food production and on situation characterized by protein shortage; market implications of this innovation.
	Upper/ Lower	Targets (feasibility of and requirements for attaining them, development and performance characteristics); organizational implications of alternative ways of attaining them.	Single cell protein on various substrates; state of the art, expected development, performance, and cost characteristics; manpower, skills, facilities, financial resources needed; required states of the art of sub-technologies.
	Lower	Individual role playing; availability of material and nonmaterial resources (including skills and capabilities)	Where is research and development pushed by individual ambition; what targets do people identify with; availability of raw materials (petroleum, etc.), of a science and technology base, of skills, financial resources, manpower, facilities, etc.; implications of location of manufacturing (e.g. in arid zones); budgets.

steam turbines, condensers, electrical generators, etc. The power industry even took on boiler technology when the latter became the most sophisticated technology in the whole functional complex with the advent of the nuclear boiler. In other branches of industry a dramatic shift from resources- or product-oriented thinking toward function-oriented thinking is going on these days, which is perhaps most conspicuous in what was originally defined as the petroleum industry and may partly become an energy industry. I have tried elsewhere to grasp the manifold manifestations and implications of this fascinating development in today's advanced-thinking industry operating in areas of rapid technological change (Jantsch, 1972 a).[1]

Operational or *tactical planning*, as we may call the interaction of the two lowest feedback loops in Figure 15, is not just about how to reach targets efficiently, but also how to set targets. The two tasks cannot be separated if business activities are understood as human, not as mechanistic, forms of action. The "autonomy of technology," the sinister correlate to "value-free science," is not a concept to be simply swallowed—it is even the least acceptable of all the deterministic concepts of which econometricians, behavioral scientists, and Marxist theoreticians keep telling us. It is least acceptable because technology is wholly a man-made artifact.

It is only through the interaction of both feedback loops in thinking which we have considered so far, that operational targets may be determined and pursued in line with a systems approach. As far as the pursuit of targets is concerned, operational planning deals with processes internal to the organization and the inputs it draws from the environment. But, as we have seen, the determination of the targets already requires a lot of consideration involving the outputs of the organization to the environment—and it is not only the traditional marketplace, but the whole social system affected by these outputs, that enters here into consideration.

But it is time to introduce a consistent (hypothetical) example, before we move on, dragging this example then further along with us. Suppose we choose food production, or more specifically non-agricultural food production for our example (see Table 7).

But how can we know that we should go into developing non-

---

1. A function-oriented approach is also implicit in the Planning-Programming-Budgeting System (PPBS) which the United States Government, as well as other national, state, and city governments are trying to make their basis for medium- and long-range planning.

agricultural food production technology at all? We assumed that this was a legitimate goal. If there is agricultural food production, what could be wrong with nonagricultural food production? Is food production not something unquestionably good in itself, so that more of it is good in itself? Not so long ago we would have considered any questioning of these assumptions as ridiculous. We knew what was good because we thought we had a solid framework of reference for making up our minds about it. Today we begin to suspect that what we assumed as given perhaps represented just our culturally conditioned reflex, one particular point of view which may no longer correspond to our current dynamic situation. If the goal is not "correct," the efficiency with which it is served turns into something bad indeed.

If we were able to determine targets in a rational way, assuming given goals, maybe we can now also try to determine goals in a rational way by assuming—what as given? Well, we shall see; our thinking tends to become fuzzy as we climb up the stairs. But we can first take a look at the feedback loop extending downward from the level of strategic goals (the food production function). Because just as we went about anchoring operational targets firmly in both directions, upward and downward—which gave us the possibility to simultaneously select targets and plan for their realization—we may now try to anchor strategic goals in both directions. Thus, what we call *strategic planning*, and what has to do with the selection, clarification, and implementation of strategic goals, again works chiefly in two feedback loops. The lower one happens to fall together with the upper one of operational planning—confusing, is it not? Why can't we neatly separate the two types of planning so that we can focus on one at a time? The answer is that we ought not to deal with them separately, that we ought to strive toward an intricately intermeshed way of thinking—or else we might always fall into the trap of keeping a goal constant and at best varying the operational targets. As targets depend on goals, so goals do on targets. We find here the same complementarity of internal and external aspects of evolutionary, hierarchy-building processes to which I pointed already in previous chapters.

When we discussed the feedback loop between the target and the goal levels (the second loop from the bottom) under the heading of operational or tactical planning, we focused on the general implications and systemic outcomes of the introduction of specific operational targets (products and services) into the systems of human life. In other words, we translated recognized opportunities into higher-level implications and tried to apply restrictions and ways of structuring the opportunities from a higher "vantage point," for which we took a more or less fixed goal. In discussing the same feedback loop now with the aim of setting and shaping a goal which is no longer considered invariant, we may emphasize the ways in which the "grassroots movements" of individual and organizational entrepreneurship contribute to giving shape to a goal which now is a societal and no longer merely an organizational goal. Food production is a societal goal, although it is served by organizational and individual activities. In particular, we may look at organizational (or instrumental) role playing and how the intricate dynamic pattern of organizational interests, potentials, and ambitions works out. We may, in our example, look at the ways in which an integral functional approach to the production and distribution of Single Cell Protein on a petroleum basis develops when the petroleum and food industries join hands—establishing an unbroken chain from petroleum prospecting, recovery and refinement through protein production, all the way to the manufacturing and marketing of finished food stuffs. We may scan and assess other ways of industrially producing not only Single Cell Protein, but also other non-agricultural food products, and we may keep an eye on promising avenues which academic research pursues in this area. We may assess to what extent and in which direction the business organizations themselves would change in playing their roles, for example becoming more consumer- and service-oriented and developing the corresponding organizational forms, striving for organizational flexibility in order to provide a variety of technological (target) responses, and so forth. But we may also look at the varying spectrum of targets, at the changing technological "decision-agenda," as we vary the formulation of our goal. Translating the implications downward to the target level in this way may stimulate quite new responses in terms of ideas for new targets, which in turn vary the organizational role playing and thus feed back into the formulation of the strategic function on which we focus.

In the upper loop of strategic planning, which is now the third from the bottom, we establish a feedback linkage between strategic goals and *policy objectives*. A policy determines the direction and momentum of the course a dynamic system takes—be the policy planned or unplanned, explicit or implicit. "Maximization of profit," "growth," "dominance," and the like are policy objectives of an older type. But here we stop and reflect—what are we talking about, a corporate policy or a policy for society? Again, the trap of separating business from society is wide open for us. The answer, of course, is that if the business corporation is to be considered as an integral part of society, the difference between the two types of policy vanishes. Corporate policy is viewed as an integral part of society's policy. The way we have climbed the stairs up to this level on the preceding pages, leaves us no other choice; strategic goals already were viewed as functions in and for society, not as indigenous to the corporation. Thus, in speaking of policy in our context, we mean the regulating principles of those societal systems to whose design, building and operation the corporate activities contribute. Concern with this type of policy is the ultimate consequence of the concept of "industry as a planner for society," on which much of the discussion so far has been tacitly based.

Thus, in the upper feedback loop of strategic planning, we inquire into the implications for social system dynamics of evolving goal patterns. In doing so, we assume for the moment that a policy is given. More specifically, we may do this by looking at *institutional entrepreneurship*, at the changing and innovating roles the institution of business considers and formulates for itself, and through which it contributes to a policy. Here, finally, we come to the right level for questioning the concept of covering the earth with red paint, or any other product. Is this the role of industry—or is it not rather participation, in a decisive role, in the design and building of the systems of human life? Is it the role of industry to pump cars into cities, or is its role rather to invent and plan for entire urban transportation of novel design which would enhance urban life instead of smothering it? Should business go on interacting with society and preying on it in fragmented ways, zeroing in on the individual consumer, appealing through the media to his subconscious ego and making him forget the concerns for his systems of life? Should business continue to hammer into people fragmented and conflicting images of the future by massive and all-pervasive intervention, which in the United States has reached or surpassed the investment in the Vietnam war when it was at its peak, and which may well have equally devastating consequences? Or should business understand its role as one of encompassing responsibility for the design of systems of human life—for the design of the social artifact of freedom?

In our food production example, we may consider here the consequences of extending food production by nonagricultural technology, what the relief of some of the food shortage pressure on the world population may mean in terms of population growth and other possible pressures on it, what it may mean to make food production partly independent from fertile agricultural land, the ecological implications, and those of bringing up the coming generation with food rich in protein, that is of a high nutritional standard. The criteria to be applied here would have to do with total system dynamics, even world dynamics, with focal aspects such as stability and resilience, survival (in itself a multifaceted, complex notion), and the long-term prospects for freedom and what we may lump under the notion of quality of life. Economic considerations would form just one aspect of these whole-systems criteria.

Thinking in these two feedback loops of strategic planning, we may then be able to formulate more sharply the strategic goals. In our example, we may inquire into the characteristics of current, and the potentials of future food production and distribution generally, what the addition of specific nonagricultural food production technologies would mean in quantitative and qualitative terms, in terms of land use, manpower, skills, funds, etc.

As typical business attitudes emerging from emphasizing one or the other direction in the lower feedback loop of strategic planning, we have already mentioned the exploitation of opportunities on the one hand, and function-orientation on the other. Both attitudes are alive in business today. What attitudes would result from stressing the types of inquiry and action emanating from thinking in the higher feedback loop of strategic planning that is the third from the bottom? What we find here, beyond organizational role playing, would really be attitudes applying to the whole institution of business rather than to individual organizations, although an institutional attitude, of course, can express itself only through the attitudes and actions of its member organizations, that is, the business corporations. But the question is no more "What can my company do in the established framework of business roles?" but rather, "What roles should business assume, and is my company equipped to enact them, or how will it have to change in order to enact them?" Such a general role has already been hinted at above— namely, the active participation of business in the design, building, and possible operation of the systems of human life—in particular systems of sociotechnological nature. We may call this the *systems design role* of business. To assume such a role appears to be the next step in an evolution which, at present, has arrived at changing business organizations to become function-oriented. It would encompass the notion, sometimes ventured today in different contexts, that business would increasingly compete in terms of ideas, plans, and generally "social software" rather than in terms of hardware. Business would become one of the central "knowledge institutions" which some sociologists expect to become the predominant type of institution in the postindustrial society, just as the production institution of industry has dominated the industrial society for more than a century now. Where the systems design role of business corresponds to an emphasis on the upward reaching arm of the feedback loop linking goals to policies, *systems regulation* would be the basic attitude corresponding to an emphasis on the downward reaching arm. We are going here beyond any institutional framework of a traditionally defined type. System regulation also cannot be the task of the institution of government alone. It will have to engage the whole of society in ways about which we can only vaguely speculate today.

There are still a lot of open questions. Where do we obtain the criteria by which we judge the "correctness" of our strategic goals? From fixed policies, which are sometimes also called religions or ideologies? How can a policy be fixed at all if it is supposed to regulate a dynamically evolving system? Can we set the system on a specific track and let it run there on and on forever? I have already argued in previous chapters that, at least in times of major transitions such as ours, policies have to be centered in the tension field of opposites; they are alive and have to be continuously reformulated.

We cannot live without rules of conduct, so we have to design and continuously redesign sets of them, that is, *norms*, the ensemble of which forms our ethical frame of reference. But, since we are dealing with whole systems of human life, it is no longer the ethics of individual behavior which matters here but the ethics of whole systems. Thus, the level of policy appears suspended between the levels of goals and norms, and *policy planning* works through the two adjacent loops.

The feedback loop which reaches downward from policies to

goals, and which falls together with the upper feedback loop of strategic planning, deals with institutional role playing and the invention of new roles. It is the interplay of institutions which forms as well as enacts policies. We would look here at the roles the institution of business defines for itself and enacts; for example, the already mentioned concern with the design and building of whole social systems rather than with the mass production and fragmented distribution of products and services. The question whether business should assume a responsibility, shared with other institutions, for the whole of society has to be posed and answered here. Certainly, in our technological age, the principle institutions responsible for the management of science and technology—industry, higher education, and government—are called upon to work together in such a way that they all contribute actively to making science and technology fruitful for the design, building, and management of human systems. This is not the case today, and we have to speculate how such an institutional pattern of role playing might look. If a policy of stability rather than growth is envisaged for the systems of mankind, it is clear that business cannot go on playing the role of boosting economic growth it has assumed so far; something will have to be done about the self-exciting nature of the economic system. This is why policy design and institutional change are inseparably linked together. In our food production example, business would have to establish its proper role as dependent upon possible world policies which may call for an increase of food production to stave off widespread famine, but which may—with regard to the consequences in population growth—also conclude that keeping a certain level of food production is the only possible way to maintain a certain level of population. The towering moral issues which are faced in designing policies cannot be demonstrated more dramatically.

These moral issues come into focus directly in the upper loop between policies and norms. Certainly, we may go along with such notions of whole systems ethics as global sharing of responsibility for world development, for the improvement of distribution (closing or narrowing down the existing and widening gaps), and so forth. But we must also face the real meaning of designing and establishing a man-made ecology which is supposed to be a viable dynamic system. Nature favors resilience over efficiency and productivity—in fact, it seems that these concepts are complementary in the sense that maximizing one implies sacrificing the other (see also Chapter 15). So far, when we consider designing a world policy at all, we think we can do better for our own species. We want individualism *and* whole system viability, we want to be efficient and productive *and* create a resilient, highly self-regulating system at the same time so that we do not become bothered by disquieting moral issues and can go on enjoying the freedom we think we can create.

Perhaps there is something more basically wrong with our concepts of institutions and man-designed systems of human life. Maybe, there is something wrong with the institution of business *per se.* Isn't it the institution specifically designed to bring efficiency and productivity to our human systems? And are we calling now on the same institution to participate in building systems of a free society, systems of high resilience which would be self-regulating and need neither dictators nor many leaders?

There seems nothing but confusion and frustration waiting for us in the dark at the top of the stairs. This may not be so different from a business corporation whose top executives earnestly try to apply a systems approach to the management of their organization. A systems approach to the questioning of our purposes and the good and evil of our actions always leads us on to confusion and frustration—and then tends to leave us there. But this is not bad, because we are now confused "at a higher level," we have switched to a new dynamic regime of thinking about the future which removes some of the confusion which used to reign at the level of day-to-day decisions. When, at the end of a two-day interaction in the framework of a top-management seminar I gave in Denmark, one of the participants complained about just that, to be now confused "at a higher level," another participant responded that in his company this elevation of confusion to higher levels had been recognized as the very gist of long-range corporate planning—a truly evolutionary perspective bringing "order through fluctuation" to corporate planning.

We did learn something, after all, in climbing our intellectual prosthesis which we thought would help us to put more order into our ways of thinking and acting. We learned at least that what appears as good at one level may turn into evil at the next higher level; that is, in the light of the purposes which emerge there more clearly; and that what made sense in a given framework may become non-sense because the framework itself may turn out to be highly questionable. As Michael (1973) points out, learning to plan implies planning to learn—we have never the full knowledge at hand which an "ideal" systems approach seems to require.

Maybe, we did learn that we ought not only to struggle upward from where we stand—somewhere in the middle of the stairs—building on what we have in our hands, not *only* I mean, but that we also ought to try to start simultaneously from the other end, questioning our ethics and attempting the design of an ethics of whole systems, to become aware of the challenge to design radically new policies for our human systems and for the whole world, and to think of the institutional roles and their modes of becoming effective which are required for designing and enacting these policies. Management is always an affair involving feedback inter-action between two or more levels. As Pattee (1973) points out for hierarchical order quite generally, the structural alternatives on one level are subordinated to a higher-level descriptive process. If we deal with self-organizing hierarchical systems, this principle appears as just another aspect of the complementarity of internal and ex-ternal factors of evolution. I shall elaborate on this most interesting question further in Chapter 14.

In this chapter, the systems approach has been presented in a rather unusual way, by speaking of levels of planning and insisting on what we may call *vertical centering*. Much more often, the systems approach is presented as a way to something like the *horizontal centering* of different attitudes about society, different points of view and criteria, kinds of disciplinary knowledge and aspirations. Of course, centering here does not mean consensus or uniformity; on the contrary, what is generally envisaged is a dialectic approach, also called horizontal management (Cleveland, 1972). What is anticipated here are roles in feedback interaction—again the systems approach does not just build an appreciated world, but also searches to regulate it. It is unimportant from which angle this interaction starts, from the system or from a multitude of partial aspects, or from one particular aspect. Thinking proceeds again in little feedback learning steps, continuously alternating between looking toward the system and looking toward partial aspects, seeking coordination between pairs or multiplets of aspects and working toward fuller integration of all system aspects.

But if the method of the systems approach is somehow built like an evolutionary system itself, as seems to be the case, there is no logical way to start with its explanation. In fact, horizontal centering of points of view and vertical integration of hierarchically ordered concepts imply each other, as we have seen in climbing the stairs from individual through organizational to institutional entrepreneurship, from simple attainable targets to issues of systems and world regulation, from individual efficiency to ethics of whole systems. Horizontal centering does not mean a market research of opinions and interests. It implies making them effective in a whole systems context, and this can only be done by ordering and coordinating them in some way. A business firm may have many products in development or on sale, a few strategies or functions it wants to contribute to, and one corporate policy—the more of its activities can be ordered in such a way, the more effectively they may be carried out.

But the systems approach, as presented in this chapter, includes at least two or three more dimensions of centering: One is a curious pair of identical twins, *time* and *causality*. This pairing is fortunate insofar as it taught us a different view of time as we changed our view of causality. Where we used to assume linear and sequential causality—A engenders B which in turn engenders C, step by step —we now see intermeshed feedback loops with complex outcomes of events occurring in them and we see causality also working backward in time, from anticipations and goals toward action in the present. What we want is always in the future, but it can guide our actions in the present. In the emerging methodology of forecasting —where the linear and sequential mode corresponds to deterministic predictions of the econometric type—we have learned to speak of exploratory and normative approaches, and we have further learned that no forecast makes sense at all which does not include both approaches in a systemic fashion (Jantsch, 1967). This is another aspect of the cybernetic nature of all design in the realm of human systems, which differs so much from technological design from which we thoughtlessly draw our contemporary stereotypes.

Again, it is unimportant from which causal view and at which time-distance the process starts. It is the intricate interweaving of both points of view that counts. If the exploratory direction dominates in forecasting, either linear planning (adaptation to and exploitation of system dynamics inherent in the present situation with its particular constellation of forces) becomes reinforced or change becomes an end in itself. If the normative direction dominates, Utopian system-states or goals are stipulated without clarification of the possibilities of pursuing them. The former deformations are characteristic of bureaucracies and economic life today, the latter of futuristic speculations. Both are useless for the full-scale planning process, but are conspicuously present in current discussions of medium- and long-range futures.

The other dimension of centering is the one which leads from forecasting (as the inventive core) through planning and decision-making to *action*. There is much feedback interaction, much learning, on the way to action, alternation between possibility and potential, input and absorption of ideas, imaginative and realistic attitude, dialectic and cybernetic modes of approaching future states. In the presentation chosen for this chapter, a most important aspect of this type of centering came out clearly: that *the selection and the implementation of targets, goals, and policies cannot be separated*. Therefore, every level of planning—operational, strategic, and policy planning—works through two feedback loops, which overlap for adjacent levels of planning. We may thus also say that the action dimension is paired with a *moral* dimension in an analogous way to the pairing of time and causality. A forecast and a plan already bear the moral burden of the decision to be taken and the action to be enacted. This teaches us to become suspicious about "straightforward" activist attitudes and to work again in feedback, that is learning loops in forecasting and planning until we have built up enough confidence that our action is moral by some ethical code—hopefully by the ethics of whole systems. It is the appreciated world which permits us to conceive and act with a moral dimension. This also implies that there are definite limits to the use of mechanistic and routine techniques in forecasting and planning, including—and perhaps more so than the simpler prostheses for our thinking—the fancy models and all the mathematics of decisionmaking, which try to make us believe that there is an amoral mechanism which makes society tick and which may be "optimized" in its functioning. Again, the systems approach elevates inquiry to the mythological level.

But we may identify still one more type of centering which is perhaps the most important one. We may call it the *centering of areas of regulation*. With Vickers (1970), we may distinguish four important areas in which regulation is breaking down in our days: economy—the relationships between our activities as producers and consumers; politics—the relationships between ourselves as doers and as "done-by"; ecology—the relationships between ourselves and the physical milieu; and appreciation—the shared basis of values, standards, and interests which forms the foundation of any living culture. As we moved up from operational targets to strategic goals, from industrial products to functions they are supposed to serve in society, the concern for political regulation began to join that for economic regulation. As we moved higher to policies, ecological regulation appeared as equally prominent with the former types—how to regulate systems so as to maintain a livable environment became the central issue. And as we moved still higher to norms, the regulation of appreciation moved into the center. And there, the upward striving of the systems approach stops. All that planning, all that systems approach cannot be done if we do not arrive at a new culture which is alive and can be built on. A new culture which will require new forms of institutions and will make them possible. A new culture which will bring new and integrated forms of regulation.

Perhaps the systems approach is after all not a bad way to get ready for a new culture. It touches almost all our serious concerns, as we have seen; it pierces crisscross through all we see deteriorate and all we would dearly wish to redesign in our human systems. If exercised, it will maybe precipitate the advent of a new culture and guide it to our hearts.

TOWARD INTER- AND TRANSEXPERIENTIAL INQUIRY: EDUCATION FOR DESIGN

So far in Western society and its forerunners, considerable emphasis has always been placed on expressing all aspects of experience in terms of rational knowledge—and to neglect to an increasing extent those aspects which resist such reduction to rational knowledge. The representation of continuously enriched experience of a complex reality through rational knowledge resulted in a "coherent trend toward optimal organization of the cognitive-semiotic modality" (Marney and Smith, 1972): from the axiomatic model of scientific thought (Greek mathematics and philosophy) through the empirical model (early modern physical science) to the conceptual model (the contemporary coalition of the formal and the experimental sciences).

Cognitive systems evolve by the same feedback interaction as do human systems. The processes of learning and design are identical. Thus, it is certainly justified to speak of a noetic evolution. This is even a most significant extension of the concept of evolution, which may be expected to have tremendous consequences for the representation and organization of knowledge. It implies that sufficiently nonequilibrium systems of information and knowledge may be expected to exhibit an internal evolutionary drive toward new regimes of noetic organization. Our task would then mainly be to manage and interpret such forms of noetic self-organization as may, indeed, be found in the evolution of great concepts in science and philosophy.

The very idea of interdisciplinarity, understood as a mode of organizing rational knowledge toward a purpose (Jantsch, 1972 a) is conducive to an evolutionary as well as a normative view. But is this all we need to focus our attention on? Is evolution in the universe, and evolution and normative design in the organization of society and of knowledge the only dynamic categories which set the stage for our discussion of human design? Should we just continue and redouble our efforts to express all our experience in knowledge equivalents and thereby restrict ourselves to a narrow notion of merely rational inquiry?

Current theories of human brain functions (summarized, for example, in Eccles, 1973) distinguish between ideational/linguistic, or rational/analytical functions, located in the left half of the brain (significantly called the "dominant hemisphere"), and stereognostic, or musical/holistic functions, located in the right half of the brain (the "minor hemisphere"). Only the rational/analytical functions are so far assumed to be in direct interaction with consciousness. This certainly is not borne out by many of the experiences encountered in the "inner way" of shaping consciousness, as discussed in Chapter 9. We have to assume that noetic evolution, the holistic grasp of new paradigms or noetic regimes, works primarily through the musical/holistic side of understanding and in both directions, making its impact on consciousness/subconsciousness directly as well as through the rational/analytical side of the brain functions.

The latter process has already found a theoretical formulation in Kuhn's (1962) theory of how scientific concepts evolve. "Normal science," the substantiation and elaboration of the great carrying concepts, the paradigms, can employ the scientific method in a narrow sense, in other words processes of induction and deduction. But the mutations in the paradigms themselves transcend the mechanisms of "normal science"; they emerge from intuitive or revelatory processes. Again, they may be pictured in terms of mutations to a new dynamic regime of organizing rational knowledge, triggered by fluctuations introduced into a noetic system whose nonequilibrium condition has become increasingly manifest. These fluctuations may hypothetically be assumed to emanate from the musical/holistic type of understanding. New scientific principles are unclear at first; they take shape in much the same way as artistic form takes shape. It is only after the formulation of a new paradigm that Darwinian processes of testing and fitting new patterns of evidence and experience take over. The evolution of new order in rational knowledge again seems to be based on the Prigogine principle of "order through fluctuation." The same holds even more plausibly for nonrational knowledge.

In a fascinating thought experiment, Laszlo (1973 a) has pictured science as a nonequilibrium system, and scientific growth as a self-regulative process of the type I discussed in Chapter 2. The evolution of science would be governed by controls which are "external to particular scientific formulations, but internal to science as an interpersonal historical-cognitive enterprise." The evolutionary change process in the system of rational knowledge would then be governed by external selection, or competition, when viewed at the level of scientific concepts, and by internal selection, or coordination, when viewed at the level of the human mind. Internal and external factors of evolution appear again as two complementary aspects of the same process; in the next chapter I shall return to this intrinsic complementarity when I discuss the evolution of human systems. Here, I want to retain the important notion that the development of science—or, rather, knowledge in a broad sense, including nonrational types of knowledge—is an aspect of an evolutionary order of process unfolding in the human mind.

Noetic evolution often follows changes in meaning, human perception and thinking as well as changes in feeling, forms of expression and communication, and values. The apparent, almost perfect, synchronization of such dramatic changes in scientific and artistic concepts as it occurred, for example, in the beginning of the twentieth century—with quantum theory and the theory of relativity, psychoanalysis, Bergson's "évolution créatrice," atonality in music, cubism and surrealism in painting, functionally oriented aesthetics in architectural and environmental design, and so forth—may be viewed as originating in preceding changes in the underlying shared experience and appreciation of the world, in "breathing air from other planets" (Stefan George): in other words, emerging from a broad-spectrum cultural change in Western society. Because of the domination of rational modes of inquiry in Western civilization, there may be little or virtually no visible phase shift between changes in knowledge and changes in other forms of human experience. But this was not always so. To many systemically oriented people, and to many of the great nonspecialized scientists, the Bible represents a holistic expression of human experience and includes knowledge of a kind which has only been approached, not attained, by modern rational scientific concepts. "Cognitive and affective aspects of meaning always go together" (Jean Piaget). Sibatani (1973) has recently stated in a powerful way a scientist's view of the necessity to take into account holistic forms of knowledge which cannot be acquired by any reductionist approach but emerge, for example, from the practice of the "inner way" as discussed in Chapter 9. It is only due to the more recent impoverishment of our experience outside knowledge, and in particular of our capability to respond to other than scientific modes of expression, that in Western society we have come to equate progress with noetic evolution and its implications in the form of a technicized world.

Still, there remains the artistic mode as an example which may be used to resuscitate the creative phase in all forms of inquiry—this is also the reason why the hopes placed in approaches which are of more artistic than scientific nature may be intuitively right. As Feyerabend (1967) points out, science has in general evaded its obligation to criticize ideology; it attempted to reestablish itself quickly and gain back, to the extent possible, its old security after being shaken up in the early twentieth century. Art has remained much more open and critical. "A critical rationalism setting itself the task of inquiring into knowledge and behavior, and improving them through critique, cannot leave aside the contributions offered by the artist."

Normative planning, underlying the systems approach as I described it in the preceding chapter, is conceived as a nonmechanistic "human action model" and brings into play the possibilities as well as obligations of man as designer of his systems. Thus, it may hopefully develop into an approach reinforcing the creative phase of inquiry. The thrust of current theory-building in planning still points in the direction of increasing "scientification," with the ill defined "policy sciences" often mistakingly given the challenge of building quantified mechanistic models of human systems. But planning, by its very nature, is dynamic, systemic in scope, and based on the feedback interaction between appreciative and creative, exploratory and normative approaches. In dealing with knowledge, it is also inter- and transdisciplinary, focusing on the organization of knowledge for the task of building human systems. But a plan includes also other elements of human experience besides knowledge; its immediate aim is not so much to convince, as to motivate. This is the reason why (as some of the foremost philosophers of science and of planning agree) a plan can often be grasped best through a story, which may be regarded as a holistic form of relating a plan, drawing on the resources of artistic expression.

Elsewhere (Jantsch, 1972 a) I have attempted to sketch the organization of rational knowledge toward a human purpose by applying the structure of a multiechelon (multilevel, multigoal, hierarchical) system (see Figure 16). Interdisciplinarity, in this

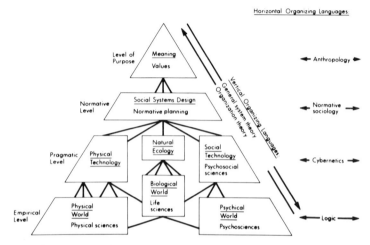

Figure 16. The organization of rational knowledge toward a purpose in a multiechelon representation (adapted from : Jantsch, 1972).

system, constitutes a mode of organization through the coordination of elements at one level from the next higher level (coordination— not goal-setting or control!), and transdisciplinarity extends this concept of organization through coordination over the entire multilevel system. Representation by a multiechelon system seems to be a viable approach to discussing any kind of organization to- ward a purpose, be it of physical, social, or noetic/spiritual nature (Mesarović et al., 1970).

Obviously, structured rationality alone—as much as it can enliven the role of the contemporary university and aid in orienting its basic functions toward a human purpose—has to be considered insufficient for design in a broader connotation. It seems worthwhile trying to extend the multiechelon systems representation to a fuller spectrum of human experience. Figure 17 attempts to sketch such a hierarchical scheme of building a human world from the basic experience of the individual through social to cultural systems. It shows some similarity to the rotary scheme of Figure 3. But what interests here is not the building blocks, but the flow of creativity entering through the individual. As basic human experience, I enter here the same categories which I discussed in Chapter 10. One level higher in the individual, they become translated into the basic values and attitudes which I listed in Table 4—for simplicity's sake, I am inserting here only the centered attitudes as they emerge at the rational, mythological, and evolutionary levels (see Table 5).

In analogy to interdisciplinarity we may now speak of *inter-experiential organization*, characterized by coordination of elements of human experience from the next higher level. Like interdisciplinarity, this type of organization is dynamic by nature, introducing a sense of direction as well as of development. Through coordination, the elements of human experience change to some extent in their concepts, substance, direction of development, and modes of expression. We may say, interexperiential organization changes the reality of human existence as we experience, build, and regulate it. It ties in with the same dialectic mode of iterative or feedback alternation between appreciative and creative phases I have identified in this book with the modes of both artistic and conceptualizing scientific inquiry. We may stipulate now that *design is inherently based on interexperiential organization*. Also implied thereby is nothing less than the recognition that design changes the reality of human existence in its total context, by changing not only the physical, social, and spiritual structures of our world but also the ways in which we experience reality, cast it into knowledge concepts as well as into concepts of what we are, feel, perceive, want, conceive, and can do—in fact, by changing not only our ways of relating to reality but also ourselves.

Transexperiential organization, in analogy to transdisciplinarity, is then called the coordination of all system elements and levels of experience—as is the case within an intact cultural system. We may say, transexperiential inquiry is a mode of organizing reason, if reason "has to do with the way in which human beings understand what human life means" (Churchman, 1968 a). The overall challenge of human systems design is implicitly the challenge of transexperiential design, involving all levels and all forms of human experience and activity. In a transexperiential context, technology is principally a way in which man relates to the world, a way of organizing broad human experience in terms of knowledge and toward a purpose.

The challenge of transexperiential approaches to the design of human systems constitutes a vast program which needs to be structured. We may ask ourselves first what kind of improvement education for design is seeking to achieve in the human "material" exposed to it. For an answer, we may come up with the following two basic notions:

—One principal aim of education is the enrichment of the substance and "effectiveness" of total human experience in the designer; in other words, the growth and centering of the designer's consciousness, themes which were discussed in Chapters 8, 9, and 10.

—The second principal aim of education is the enhancement of the designer's capability to organize his total experience toward a purpose through inter- and transexperiential modes of inquiry.

These two aims correspond to different approaches to education for design. One focuses on the substance at the four levels in Figure 17—an approach which the university has traditionally taken in the realm of rational knowledge at the two lowest levels. The other focuses on the three steps of interexperiential organization between the system levels. There can be no doubt that the latter approach is the more important today, since it enforces a dynamic attitude toward experience and implies an orientation toward a purpose, and thus an orientation toward design. The same stance is taken by Morgenstern (1972 a) when he warns that "'we should never pretend as I fear we do too often, expecially in the introductory textbooks, that we can give systematic knowledge to our students" and quotes Paul Valéry who said that educating means "to prepare the young for situations that have never been." These are refreshing words in an environment in which management education is still largely based on the retrospective Harvard Case Method, and skills are developed to master situations of the past.

Of course, in "bridging" two levels, the substance at these levels may be taken into account as well but with a view to its "valency" for interexperiential organization toward a higher-level concept. In general, this substance will have to be included to an even more prominent extent than would be required for rational knowledge alone. A great deal has been done over the centuries to develop science and technology, and very much of it is constantly present to us, experienced by us, whereas most of the rest of human experience has become "buried beneath the mythology of our culture" (Vickers, 1970).

In this chapter, I shall consider three basic structures for education

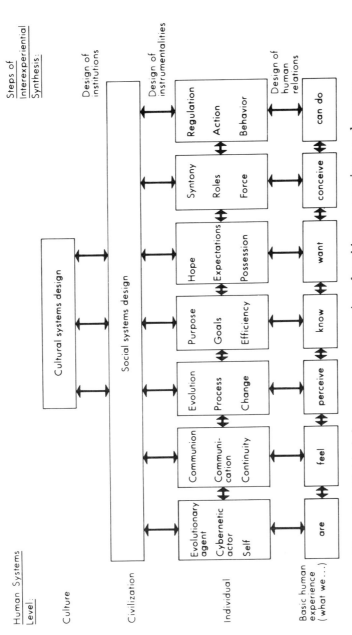

Figure 17. Multiechelon systems representation of total human experience and purposeful activity.

geared to the idea of design. These structures focus on the dynamic development of relations, or role playing, expressing the concept of human action which dominates the mythological level of inquiry. The three basic structures are:

—the design of human relations, in a general (not just interpersonal) sense, as a mode of organizing basic human experience into a web of relations with the surrounding world;
—the design of human instrumentalities and their role patterns as a mode of organizing the elements of man's world ranging from the natural and technological physical milieu to people into social systems or civilization;
—and the design of human institutions and their role patterns as a mode of organizing social systems into a culture.

Figure 18 depicts graphically how these three structures are supposed to bridge the system levels of Figure 17. In my earlier treatise on a type of university geared to rational knowledge (Jantsch, 1972 a), I have proposed a corresponding triplicate structure of discipline-oriented departments, function-oriented departments, and systems laboratories.

In addressing *education for the design of human relations,* I mean the general ways in which man relates primarily to physical and social, but also to spiritual reality. This part of education might be viewed as beneficial to all people, not just future planners and designers.

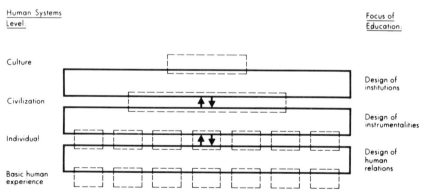

Figure 18. The focus of interexperiential education and its structures (processes) in relation to the multiechelon system of total human experience and purposeful activity (Figure 17).

Communication and creativity have hardly found a place yet in higher education. Creativity, under some heading, is taught in management, organizational psychology, architecture, the fine arts— but it is hardly developed in a systematic way, as a means of relating to the whole world. And less even is done for "communicative skills."[1] Some of the approaches I discussed in Chapter 9, which are chiefly developed by special "growth centers," are gradually finding their way into the school system from primary schools to universities. But they are in no way integrated into the overall educational process.

In the realm of iconological systems, or human relations, we may recognize as a principal approach the *conceptual method* (in science as well as in art and other forms of interexperiential synthesis). Education geared to the development of human relations would have to mix formalized and nonformalized learning (living) through novel approaches. The trend among American students to acquire "living experience" in between their formal studies, e.g., by working in the sore spots of the social fabric, and their demands for the acknowledgement of such living experience as part of the formal requirements of higher education seem to point in the right direction.

Among the more concretely rational topics which may, to some extent, be "taught" in a more formal way, of particular importance are those which emphasize the dynamic, organizing, and integrating aspects of human relations with the world. They may include, in no particular order:

—Human and environmental biology;
—Psychology and psychoanalysis;
—Sociology;
—Behavioral science;
—Critique of empiricism;
—Theory of reflective consciousness;
—Theory of creativity;
—Principles of ecosystems;

1. In the Spring Quarter of 1973, I led an experimental seminar at the University of California, Berkeley, which was devoted to the exploration of modes of human communication not normally discussed in a university—from acupuncture to telepathy and telekinesis, from plant communication to psychic healing, from psychotherapy to the techniques of self-attunement experienced by the seminar at a weekend workshop given by the Arica Institute. Many participants considered this approach as one of very high immediacy and relevance to their intellectual formation and to their lives in general and asked for more seminars of this type.

—History;
—Formation of myths;
—Cultural anthropology;
—Semiotics;
—Origin of the arts;
—Forecasting methodology;
—Aesthetics (from the point of view of natural and anthropomorphic forms and modes of expression);
—Interface of human biology, psychology, and aesthetics;
—Brain and mind functions;
—Communication theory and basic notions of cybernetics;
—Other forms of communication (parapsychology, plant communication, shamanism, etc.);
—and so forth.

But the backbone of education for the enhancement of human relations in the world may well be provided by the approaches of the "inner way" which I attempted to survey and structure in Chapter 9.

Education for the design of human instrumentalities addresses the crucial interexperiential step in our struggle for meaningful design today. It is not sufficient to set norms and develop approaches to devising social systems design which satisfy them. The central theme here is the design of human instrumentalities, of organizations of people, together with their technologies, communication systems, ambitions, expectations, goals, feelings, and so forth. On the basis of constructive elements available in human beings as well as in the surrounding world, these instrumentalities develop their proper roles and, in interaction with each other, build and manage social systems. The focus of design is thus on the pattern of instrumental (or, in some cases, individual) role playing rather than on the structure of the system in question. Of course, the design of a role pattern will usually imply some idea for an initial structural design of the system and the "positions" of role players as well as the "rules of the game."

Education, focusing on this second interexperiential step and reaching up to the normative level, will center on the key notion of a *normative method* yet to be fully developed. Such a method would attempt to systematize the inquiry into behavioral and structural norms expressed through instrumental (individual, group, or organizational) role players; it would attempt to translate human and institutional values into behavioral patterns regulated by ethical norms; and it would set a framework for the development of alternative instrumental roles by inquiring into the capabilities, goals, and modes of expression of individuals, groups, or human organizations, viewed along with their psychological, technological, anthropological, aesthetic, and social attitudes. The normative method would thus focus on man *together* with his entire world context (including, in particular, also technology)—not in isolation from it. In the realm of rational knowledge, I follow Marney and Smith (1971) in their call for innovation in the codification of alternative techniques of systems analysis based on objective-predictive and normative-prescriptive modes of inquiry, a development which has already started in forecasting and planning methodology.

Although the involvement with real-life instrumentalities and systems and their design appears most essential to the educational tasks in this domain, a few topics may be enumerated which can be "taught" in a more formal way, again in no particular order:

—Innovation processes in the social, political, and technological areas;
—Philosophy of rational creative (nondeterministic) planning and action;
—Political science (from a normative point of view);
—Policy science;
—Advocacy, persuasion, motivation, self-motivation, coercion, etc.;
—Systems approach to planning;
—Organization theory (with special focus on intra- and interorganizational flexibility, and modes of decentralizing initiative);
—Principles of complex dynamic systems (especially social systems);
—General system theory;
—Theory of dynamic nonequilibrium systems;
—Operations research and general approaches to systems analysis;
—Forecasting methodology and theory of strategic planning;
—Social indicators and general experiential indicators;
—Concepts of social systems effectiveness and cost/benefit;
—Concepts of applied behavioral sciences;
—Decision theory, decision and heuristic models, investment in dynamic situations, etc.;
—Decision-making processes, including nonrational ones (e.g., rites, magic);
—Rational approaches to resource allocation;
—Science, technology, and public policy;
—and so forth.

In education for the design of human institutions and their role patterns, finally, learning is essentially geared to a better understanding or experiencing of the evolution of values. As we simulate the dynamic outcomes of bringing into play various value patterns— through human institutions and their role playing—or, in other words, the outcomes of various policies for social systems, we try to match them with the purposes embedded in a cultural design. As new patterns of values emerge from this feedback process and are tested against new dynamic designs, we improve our understanding of their effect in role playing.

The key notion of interexperiential synthesis at this step is an *axiological method*, or value inquiry, which is yet to be built. It should be conceived so as to systematize inquiry about the nature of values represented by various institutional role players, relating them to their behavior, forecasting their implications in an exploratory as well as normative sense, and bringing them into play effectively.

The axiological method would focus on the issue of policy design, or the feedback cycle of organization and regulation which I shall discuss in the following chapter. The ethics of whole systems, which emerged as a norm for the spiritual domain in Chapter 8, may become one of the key notions for an axiological method. It has the potential of accelerating and guiding cultural change as well as providing a driving force for the design of new cultural images. Guidance is here the more valid, the more the notion of whole system ethics is based on *li*, the ultimate perspective of an universal order concurring with the order of human action. Taking the ethics of whole systems seriously implies facing moral issues of such dimensions as we normally avoid discussing or even recognizing today. We have only to think of the dilemma into which we get at the global level through the population explosion, the simultaneous demands for boosting food production and industrialization on the one hand, and for environmental and resource conservation on the other, and a variety of other imperative demands running counter to each other. Any way of dealing with these demands will raise vast moral issues—but so will not dealing with them.

The essence of "pedagogy" in this domain is, of course, practical design work in system laboratories, relating the phenomenology of total dynamic behavior of systems to their spatial and temporal morphology, but also—and above all—to the value patterns in action and expressed through the roles which human institutions assume and play. The aim is a better understanding of, or also a better feeling for, the interdependence of culture and social systems through policies, and the possibilities and feasibilities of designing cultures and social systems in a feedback process.

Some of the more clearly definable rational topics which enter here may be considered to include:

—Anthropology of the industrial and postindustrial society;
—Values and value dynamics;
—Ethics, especially ethics of whole systems;
—Philosophy of self-renewal;
—Evolutionary concepts and trends (cosmological, physical, biological, social, noetic, general experiential and value evolution);
—System theory (general system theory, feedback systems, hierarchical systems, etc.);
—Feedback and control theory, cybernetics in general;
—Information theory;
—Dynamic modeling (system dynamics, etc.);
—Theory of normative planning;
—Institutions of industrial and postindustrial society and their changing role patterns;
—Religions and ideologies;
—Holistic interpretation of great art, religions and philosophical systems;
—Theory of mind;
—and so forth.

The principal approaches to interexperiential synthesis at these three steps provide us also with key notions for learning and synthesis, as summarized in Table 8. We find a triplicate feedback interaction between loops built on the common principle of cybernetic

TABLE 8

*Focal notions for approaches to learning and synthesis at
the three steps of interexperiential organization.*

INTEREXPERIENTIAL STEP	APPROACH TO INTEREXPERIENTIAL SYNTHESIS	METHOD
Design of Human Institutions	Cultural Cybernetics	Axiological
Design of Human Instrumentalities	Social Cybernetics	Normative
Design of Human Relations (Iconological Systems)	Human Cybernetics	Conceptual

self-organization through role playing. Human relations play such roles in conceptualizing attitudes and action on the basis of human experience; in this process which I have discussed in more detail in Chapter 8, and which may be labeled *human cybernetics*, reality reverts to the basic experience and to a certain extent changes it, whereby, in turn, the role playing pattern of human relations changes. In scientific terms, this corresponds to the familiar conceptual method. The same basic process goes on at the step of human instrumentalities, where we may call it *social cybernetics*, and where the individual and the tentative design of social systems are interlinked and change in a process kept in motion by the role playing of human instrumentalities. In organizational terms, we may hope here for a normative method which has yet to be perfected. And finally, a process which may be called *cultural cybernetics* characterizes the design of cultures in a feedback loop with social systems and their policies, animated by the role playing of human institutions. The corresponding methodological approach may be called an axiological method.

The overall task of designing human systems may again be viewed as focusing mainly on the design of processes rather than structures, as has already become clear in Chapter 4. The methods to be applied are therefore cybernetic rather than heuristic.[1] Of course, the design of such feedback processes in the physical, social, and cultural sphere involves the design of structures insofar as an initial system state and an initial role playing pattern have to be introduced, if they are not given by what exists—but none of these structures is supposed to remain fixed. It must be kept in mind that the structures of which I am speaking here are primarily structures of total human experience, of values and attitudes at various levels, and may be coupled to physical or administrative structures only incidentally.

The proposed perspective of design entails enormous consequences for methodological development in forecasting—which has already found a formulation in cybernetic terms as evolving in an open loop of exploratory and normative modes (Jantsch, 1967)—in planning, policy design and analysis, political science, law, economics, sociology, history, etc., but even more so in interdisciplinary

---

1. I wish to emphasize here again that the notion of interexperiential synthesis as I develop it here, is geared to the organization of role playing as primary change process. However, role playing—a notion located at the mythological level of inquiry—is but one change process to be taken into account, although it makes the fullest use of basic human experience. In the following chapter I shall also discuss other change processes.

approaches to human systems. Some branches of psychology, anthropology, and linguistics are already well advanced in exploring the cybernetic perspective mainly in the realm of what has been labeled above, human cybernetics. The challenge of social cybernetics is finding the response of some of the best interdisciplinary minds today. But there is little awareness yet of the potentials and the necessity of cultural cybernetics.

As the backbone needed today most urgently for a total and fully transexperiential approach to design, one may conceive a *comprehensive normative theory* which has yet to be developed. Marney and Smith (1971) sketch the elements of such a normative theory for the domain of rational knowledge only. A normative approach to total human experience would extend these rational concepts perhaps to embrace the following notions:

—*Formal criteria* will be concerned with anthropomorphic form and measure;
—*Empirical criteria* will be concerned with comprehensiveness of human experience involved;
—*Pragmatic criteria* will be concerned with inter- and transexperiential coordinability;
—*Aesthetic criteria* will be concerned with systemicity (degree of being "tuned in" to social and evolutionary dynamics) in the organization of total human experience;
—*Ethical criteria* will be concerned with the improvement of whole human systems;
—*Evolutionary criteria* will be concerned with a sufficient degree of non-equilibrium (flexibility in organization), meliorative trend, and viability in the light of "order through fluctuation."

Most of what I have formulated earlier (Jantsch, 1972 a) on the general characteristics and aims of a process of higher education geared to inter- and transdisciplinary inquiry remains valid in the extended framework of education for design, and even more so. The interexperiential learning processes cannot be the subject of teaching in the traditional sense, nor of training in the use of methods. The learning processes in school and in real life have to become identical, they have to be experienced themselves in order to become part of human experience. Whether the instrumental form in which this can be achieved best will be called a school or a university or something else, and whether it will even faintly resemble contemporary educational instrumentalities, is of secondary importance. In any case, such an instrumentality will be part of human and social life—it will be political and an instrumental role player itself. And education of this scope in general will certainly play an institutional role of considerable importance in our search for viable cultural designs.

As in the transdisciplinary university, students may focus on one of the proposed educational structures only, that is on one step of interexperiential synthesis, or may go through two or all three of them. They may "fluctuate" between them, for example, experimenting with the design of institutions and policies, and interspersing this with studies in the design of instrumentalities or physical elements, such as technology. Industrial designers of the traditional type may stay within the "human relations" section, whereas architects, city planners, and social planners may focus on the "instrumentalities" step, and some of them on the "institutions" step. For all of them, the proposed approach to education may offer a broader basis of human experience and enhanced capabilities to apply it to design.

What is missing altogether in the present picture is the cultural designer, focusing on human institutions and their roles. Yet without him, even the best-motivated and most sophisticated efforts of interexperiential design will be rendered futile. Only transexperiential synthesis, the coordination of total human experience and all forms of human systems, is capable of providing that sense of purpose and meaning which characterizes a viable culture. As I have already stated, design is inherently a transexperiential challenge— if we fail to develop our responses to this challenge, above all in education, human experience and human systems will become ever more fragmented and further removed from a reality which appears to be growing increasingly inhuman.

But role playing or self-organization based on human *action*, is only one change mechanism at work in the human world. It is the one of which we can become most conscious and which, to a certain extent, is the aim of education in a more traditional sense. In the next chapter, I shall try to take a broader look at the self-organization of human systems, embracing evolutionary perspectives beyond human action. Again, the principle of "order through fluctuation" will shed new light on processes of self-organization.

PART V

Reality Evolving

The presence of the Spirit:
It cannot be surmised,
How may it be ignored!
From the ancient Chinese Book of Songs

THE reality of the human world may be viewed as evolving through successive "waves" of organization—from ecological through social to cultural and further to psychic organization. These "waves" constitute macromutations to new dynamic regimes in the evolution of mankind. They form a hierarchy bonded by the interaction of internal and external factors of evolution. What may be understood as an external (competitive, Darwinian) factor on one level, appears as an internal (coordinative) factor at the next higher level. In this way, a hierarchy of human systems unfolds from the individual to mankind and further on to noogenesis, the spiritual aspect of evolution. The vague contours of an emerging order of process in the evolution of the human world permits the recognition of a few principles which may govern the near future of mankind—and thus also the sensitive and creative design of policies and perhaps future cultures. In the nondualistic blending of its concern for stability and its openness for attraction toward new order (and thus instability)—of planning and love—human design now appears itself as a process expressing and enacting the evolutionary urge toward an ever renewed possibility for energy exchange—toward life in a broad and all-pervasive sense.

(14)

POLICY DESIGN—
ORGANIZATION AND REGULATION

Policy design is the process characterized by the forward arrow of organization, reaching out from the malleable appreciated world to reality, and by the arrow of regulation, transmitting the "response" of the real world, which is hard in comparison with the soft, appreciated world. By organization, I shall in the following understand any process of establishing order, whether we are conscious of it or not.

Evolution in all spheres may be characterized by particular patterns of interplay between internal and external factors of evolution (see Chapter 3). External selection works through competition, internal selection through coordination (Whyte, 1965). In Chapter 3 I mentioned that internal factors are now receiving increasing attention, after evolutionary theories from Darwin to Teilhard de Chardin had recognized exclusively external factors. In fact, understanding internal selection leads to a much better understanding of the evolutionary process itself (Thom, 1972; Prigogine et al., 1972). Internal factors generally come into play first, ensuring internal consistency of individual ontogeny with the overall conditions of morphogenesis, which only in the past few years we have started to see in the physical and biological domains. Internally successful mutations are then further screened by external selection in interaction with the environment. Possibly, this interplay is to be understood less as a sequence than as an intricate feedback loop, with return and reform mutations induced anywhere in the process, much more so than it may appear to us now.

However, as Whyte (1965) already points out for the biological domain, new species generally start their phylogeny by finding and occupying a new ecological niche, or adaptive zone. In the early phases of phylogeny, external selection may then be expected to play a relatively minor role in comparison with internal selection. As

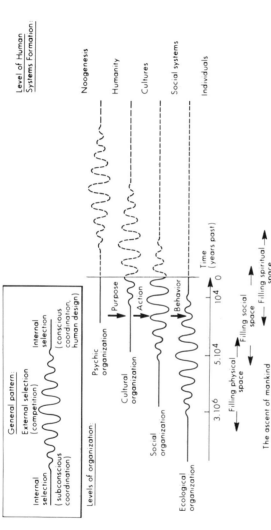

Figure 19. The hierarchical relationship of waves of organization in the evolution of mankind. Each wave passes through three consecutive phases which are characterized by: 1. Predominance of subconscious internal (coordinative) factors; 2. Predominance of external competition, or Darwinian processes; and 3. Predominance of conscious internal (coordinative) factors, expressed through human design. Organization does not "jump" from one wave to another but becomes more richly orchestrated as new modes of organization come into play, take the lead, and subordinate themselves. This progression may also be understood as the evolutionary transmutation of mankind's energy along the seven classical Hindu *chakras* : 1 = physical survival; 2 = sensual pleasure, sex; 3 = power; 4 = heart, love for all mankind; 5 = in search of God; 6 = wisdom; 7 = ultimate union.

the niche becomes more crowded, external selection, by the "survival of the fittest" principle, plays an increasingly decisive role. In the human domain, the filling of the physical space—an ecological niche—formed just the prelude to a phase of the filling of the social space, a development which is now nearing its end (see Chapter 3); and filling the spiritual space may yet lead into another phase dawning at mankind's horizon. We may speak of successive waves of organization, each showing different characteristics.

In Figure 19, I tried to sketch in a highly simplified way the interaction between these phases of mankind's evolution, or waves of organization, as I view them. They are governed by a hierarchical relationship which also expresses policy design at different levels of human systems. But it also lets the notions of internal and external factors of evolution appear as relative to specific hierarchical levels. For the most active time of each organization wave, the following hierarchical scheme may be drawn:

$$
\begin{array}{lll}
\text{Psychic organization is} & \left\{\begin{array}{l}\text{internal to}\\\text{external to}\end{array}\right\} & \text{noogenesis} \\
& & \text{humanity} \\
\text{Cultural organization is} & \left\{\begin{array}{l}\text{internal to}\\\text{external to}\end{array}\right\} & \\
& & \text{cultures} \\
\text{Social organization is} & \left\{\begin{array}{l}\text{internal to}\\\text{external to}\end{array}\right\} & \\
& & \text{social systems} \\
\text{Ecological organization is} & \left\{\begin{array}{l}\text{internal to}\\\text{external to}\end{array}\right\} & \\
\text{Biological organization is} & \text{internal to} & \text{individuals}
\end{array}
$$

The lower end may be readily extended in the same way to cellular and molecular organization. What appears as internal factor at one level, constitutes an external factor at the next lower level; both are but different aspects of the same process of organization, a process which always links two levels in the hierarchy of human systems. The competition for physical survival, for example, appears as a Darwinian, external struggle viewed from the level of the individual; but at the level of social systems—of tribes, communities, or peer groups—it appears as a process fostering coordination by the development of role patterns, task sharing, and mutual expectations.

This intricate interaction, or even complementarity, of internal and external factors of evolution may be viewed as an instance of a profound principle of hierarchy theory to which Pattee (1973) points: the structural alternatives on one level are always subordinated to a higher-level descriptive process. But I suspect that, quite generally, structural and descriptive views are but complementary aspects of the same process. The interaction between structural and descriptive levels which, in bioorganisms, is established by the genetic code, generates the dynamics of self-organization. Internal factors of evolution would then become expressed through description in a broad sense (as biologists, for example, speak of the cell's genetic description of itself), and external factors through structural alternatives which are still open.

How does human design fit this hierarchy? A comparison of Figure 19 with Figure 13 (see p. 122) shows that each level of human system design interacts with two adjacent levels of evolutionary organization processes, so that the above representation may be simply inverted in the following way:

Cultural system design is	{ internal to { external to }	cultural organization
Social system design is	{ internal to { external to }	social organization
Physical (individual) system design is	{ internal to { external to	ecological organization
		biological organization

In this way, the design processes become integral aspects of the hierarchy-building evolutionary processes. *The ultimate task of human design may thus be recognized in the enhancement and reinforcement of the hierarchical bonding between the levels of evolutionary organization.* The interaction between interference and transformer systems (see Figure 3, p. 52) also appears now as an expression of the intricate connection between design and evolution in human systems.

It also becomes clear now that individualism, as it reigns in Western culture, cannot be integrated "upwards" and patched together to form an ethics of whole systems. Vickers (1973 a) justifiably calls the present antithesis between our development as members of institutions, and our development as persons "the most pathological symptom of our age." And Böhler (1973) argues incisively that the present growth syndrome is geared to individualism, that the constantly evoked notion of human rights reflects individualistic claims. The futility of the attempts to preserve individualism and build on it a viable social order, Böhler argues, leads to ideologies, to *ad hoc* myths. The individualistic claims of competition, democracy, and nationalism become embedded in the corresponding myths of market economy, equity, and integration (world state, eternal peace). But individualism is served by ecological organization—without a mutation to social organization, to a higher step in the hierarchy, it will be impossible to come to a meaningful design of social systems.

The evolution of a higher wave of organization in the human domain permits the internalization of the next lower level of organization through increasing consciousness and applicability of human design. Each wave of organization evolves from a subconscious phase in which objective internal factors govern the morphogenesis of organizational forms in the direction of increasing consciousness and the domination of internal factors introduced by human design, which, in itself, is a combination of subjective and objective internal factors. On the way to the realization of the human design potential at a particular level, the corresponding wave of organization passes through a phase of dominant external selection by way of competition.

We may also say that each kind of organization is characteristic of a particular dynamic regime of human life. Thus, the succession of waves of organization may be understood as another manifestation, at a macrolevel, of the basic evolutionary mechanism of "order through fluctuation," of a rhythm alternating between stabilization and mutation, as it has been described for nonequilibrium physical systems (Glansdorff and Prigogine, 1971). The highly competitive phase, or externalization, of each wave of organization, would then introduce the fluctuations which give rise to mutations. Each new regime is characterized by a new complexity which, in turn, may be understood in terms of equivalent energy or information. A new regime first acts out complexity in terms of energy, producing considerable entropy, and gradually moves toward a phase in which information comes into play. In such a way, ecological organization moved from the energy-utilizing concept of hunting to the information-utilizing concept of agriculture; social organization from political power and legitimacy to advice-giving, analysis, and flexible management; cultural organization is in the process of moving from closed religious and ideological power systems to open dynamic regulation; and psychic organization may perhaps be seen as developing from alchemy, or the utilization of psychic power to effect change, toward subtle forms of regulating evolutionary forces.

The waves of organization may also be understood as evolution of human consciousness, or as the transmutation of human energy, which may be described in terms of the seven Indian chakras I mentioned in Chapter 9 and briefly outlined there. Ecological organization has to do primarily with the lowest, the first chakra, namely survival or conservation in a physical sense. Social organization brought the second and the third chakra, sensual pleasure and power, into focus. Cultural organization brings in the heart chakra, love as the basis of syntony among people. In spite of the Christian message of love nearly two thousand years ago, person-oriented love looks only like a prelude in this broader view, like a reflection of a cultural category as it appears at the lower echelon of social organization. The incipient psychic form of organization, and a subjectively lived out culture—as great artists have anticipated it for a long time—will then lead mankind on to the fifth chakra, the search for God or a new transreligious mysticism as it already seems to be springing up in an increasing number of psychically sensitive people. Thus, the level of interaction of the specifically human energy may come to be transmuted toward the highest chakras, wisdom and ultimate union.

In this discussion, human design appears as an internal factor, as a conscious form of internal coordination. As physical space became filled and the ecological organization of mankind primarily determined by Darwinian, external, principles, social organization—the development of social roles and the filling of social space— introduced and guided the human design of man's ecological relations with his environment, thereby increasing coordination and decreasing competition for survival. With highly developed social organization, man's relationships with his physical milieu became increasingly planned, or in other words subject to internal, conscious, coordinative processes. At the same time, competition for social space increased and external factors such as the market mechanism or struggles for power took over in social organization from the early subconscious coordinative factors which gave rise to the formation of families, tribes, and villages.

It is interesting to compare organization in the human domain to organization in animal societies. In beehives, for example, the standard form of stable social system is characterized by the presence of only one queen on which rests the entire female burden of procreation. The diffusion of an enzyme, a chemical signal, indicating the presence of a queen in the hive, regulates the birth of bees.

As long as the presence of a queen is signaled, no other queen is supposed to be born. But errors, random fluctuations, are of course not excluded. It appears that these errors increase sharply when the beehive exceeds a certain size, or level of population. Before long, a second queen seems bound to appear and the swarm breaks up into two parts, one of which emigrates and builds a new social system elsewhere. In general, one may say that a hierarchically superior level of organization governs the conditions for the occurrence of fluctuations (which, in themselves, may be random) and thus induces a mutation—in this case, the multiplication of a social system at the same level of dynamic regime.

But in the human domain, at least after the passage of some very early phases of social organization, stable social systems never seem to have been an ultimate goal. Apparently, human social systems have grown in size and scope throughout mankind's psychosocial evolution. In Chapter 3 I outlined the regularity which Calhoun (1971) believes to have found: with rising world population, social systems did not split apart—on the contrary, with every doubling of the number of people on earth, seven units (or "social brains") joined to form a new one. In our time, we may observe many aspects of this formation of new and larger social systems: cities incorporating suburbs or growing together; the "optimal" size of the economic system growing from cities and fractionated states in the nineteenth century to national economies and further to configurations such as the Common Market or Comecon; or the formation of military and political alliances and power blocks.

Unlike animal societies which (at least viewed against a time scale corresponding to human history) tend to maintain social systems at a given level of dynamic regime, human societies definitely mutate toward higher (more complex) levels of dynamic regime. This certainly constitutes an important aspect of man's unique role on this planet. This uniqueness, again, seems to be rooted in the capacity for human design, in other words, for the conscious management of dynamic regimes. As human design learns to manage mutating regimes at one level of organization, the primary fluctuations (expressing the externalization of organization) move to the next higher level, to the next wave of organization. By virtue of this hierarchical ascent of the guiding type of organization, we are given the chance to *design, and not just suffer, mutations*—so far at the levels of ecological and social organization. We can use information which streams from the less conscious higher levels. For example, with changes becoming felt in our value systems— that is to say, with fluctuations at the level of cultural organization— we can attempt to design new codes of social ethics, new social institutions and their role patterns, or new social systems altogether. We can also design new approaches to organizing food production and other vital needs and to maintaining a viable partnership with our environment at the ecological level.

Perhaps the image is meaningful here that there is a "downward" action from higher to lower levels of organization, and an "upward" action in the opposite direction. The "downward" action would induce fluctuations in the lower levels which lead to externalization, or increasing competition, until human design understands to internalize them. But this conscious internalization builds the solid grounds for "carrying" fluctuations at higher levels; thus, the next wave of marcro-fluctuations is also induced by "upward" action. Self-organization thus appears as a nondualistic concept. The vertical law of organization is inextricably linked to the horizontal law of interaction, as Laszlo (1972 a) generally stipulates for natural systems—self-stabilization and self-organization become integral expressions of self-realization.

Social organization induces *behavior* at the ecological level. In the beehive subconscious and internal social organization alone governs behavior, or the fluctuations at the level of ecological organization. Cultural organization induces moral *action* at the social level. This type of fluctuation, at the social level, is characteristic for current forms of human society. But the impending filling of social space, or the inevitable approach toward world unity in social and political terms, sets the stage for major fluctuations at the level of cultural organization. These fluctuations, in turn, seem to be induced and guided from a higher level, namely the level of psychic organization which induces *purpose and meaning*, and with them a sense of direction, at the cultural level.

We again find in these notions the three levels of planning, or human design generally: the "know-how" of operational or tactical planning, corresponding to behavior and to a mechanistic type of system; the "know-what" of strategic planning, corresponding to moral action and to an adaptive type of system; and the "know-where-to" of policy planning, corresponding to purpose and direction and to an inventive type of system (see Chapter 12). It becomes

utterly clear now why no behavioral model can do justice to the
present phase of human evolution. Where a behavioral view represents only one level, the lowest one, the organization of human
systems is at present orchestrated at least at three levels, and human
design is consciously at work at two levels.

We gain now also a new perspective of the notions of equilibrium
and nonequilibrium. In our days, the discussion on growth and its
limitations (Forrester, 1971; Meadows et al., 1972) focuses on the
establishment of equilibrium conditions at the level of ecological
organization—population, food, nonrenewable resources, biosphere
(pollution). To the extent that social organization enters at all, the
underlying image of equilibrium would favor the "freezing" of
social evolution at the stage which it has reached in our time. The
possibility of qualitative mutations to new regimes is not recognized
by this narrow and overly rational view which accommodates merely
quantitative considerations. Even the mechanism of the beehive
seems unacceptable for the human world in such a view.

Such a discussion is not only unrealistic—because there is no
way of establishing the type of world dictatorship required to implement the prescriptions—it also epitomizes an idea of equilibrium
which is the equivalent of social and cultural stagnation, or death.
Not the appeals to exercise self-restraint are mistaken, but the implicit expectation that evolution may come to an end, or be stopped,
once we have learned to apply human design to ecological and social
organization. This is a reaction of fear as well as an expression of
hubris.

In the view taken by this book, it is the very nonequilibrium state
of mankind which induces mutations toward new regimes. The
decisive mutations to come, and to be encouraged, will pertain to
the level of cultural organization. The necessary social mutations
toward world unity will have to be "pulled" from this level. But
it is up to us to regulate our affairs in tune with evolutionary forces,
or to be squashed by them. Ecological and social balance, and the
conditions for effective human design at these levels, can only
emerge from a nonequilibrium system of mankind, alive at all levels
of organization and fluctuating at the cultural level. Any stipulation
of a static order—or the abdication of design by enforced stagnation
—will only cause greater birth pains, greater turmoil and suffering.

Man's rational hubris is very close to turning massively against him
—we are certainly becoming aware of that. But the battle for the
evolution of mankind will go on and increasingly switch to the
cultural level.

With the externalization of cultural organization, and an increasing
awareness for cultural pluralism, one might also expect the restructuring of the relations between civilization and culture, a *growing
emancipation of culture from social affairs*. This would have vast
consequences. So far, culture and civilization has come as a tightly
bonded pair, as a "package," as witnessed in the process of Western
technology becoming introduced to the Third World. Invariably,
this process also implied the introduction of Western cultural notions,
Western values, forms of government, styles of life, artistic expressions, and so forth. Aid given to the Third World turned partly into
a devastating process of softening up and washing away indigenous
cultural expressions and patterns, together with the basis of self-
identification which is so important for building relations to the
world—so important for human design.

As social organization is increasingly becoming a matter of design,
cultural pluralism may be expected to flourish *within* societies and
specific patterns of civilization. World unity cannot possibly mean
cultural unity—not yet, at least. The Western culture, transplanted
into all parts of the world and diluted over the whole globe, will
become just one of several substrates to nourish the future growth
of a variety of cultural phylae.

In the context of the precarious phase of transition from predominantly social to cultural organization in which we find ourselves at
present, it is interesting to note that Teilhard de Chardin already had
speculated that with rising consciousness suffering may also rise. He
even allowed for the possibility that there would be a final great
schism, with part of the noosphere being attracted toward God, and
another part toward the evil. But perhaps, if we view noogenesis as
a mutatory process creating the same "order through fluctuation"
which may be recognized in microaspects of evolution, a schism may
be expected to mark each phase of transition from one organizational
level to the next higher? Will the Western world then, the "high-
technology phylum" which, by so many tokens, seems to be out of
step with evolution, be the part which is going to hell? Or will a
new turn, a radical cultural fluctuation, bring it to the forefront to
become the spearhead of a more sharply focused departure into cultural pluralism? North America, the spearhead of the "high-technol-

ogy phylum," is already turning into a laboratory for new life styles, for cultural pluralism unequaled today in other parts of the world. As the standardized, uniform "American way of life" is spreading over the whole world, a kind of self-renewal already seems to be under way. Also, spiritual people from all over the world feel increasingly attracted to North America, where they sense a new spiritual force. Not the technological humdrum, not the hedonist movements—a true spiritual renaissance. Could it really be that life breaks out with greatest power where it has become a mockery? That the society most advanced in the application of a mechanistic and materialistic everyday-life credo should become the beet in which the flower of a new spiritual consciousness blooms first? But this is how evolution, how the micro- and macroprocesses of "order through fluctuation" seem to work: where the conditions for creative life, for the production of entropy, become stifled, fluctuations are most likely to lead to a mutation, to a new set of conditions which restores the chances of life.

In the present, we find ourselves close to very significant turning points in the graph sketched in Figure 19: Ecological organization has become largely determined by conscious design; but it is at best uncertain, whether it will stay this way, or whether we will move again toward increasing competition for physical survival on the planet (which might be regarded as an "evolutionary failure" due to the wrong social and cultural organization). Social space is nearly filled and social organization is characterized by a high degree of competition. However, the pendulum is already swinging back to coordination, now on the basis of human design, of social planning with its core ideas of equity, equal opportunity, and an assured minimum material standard of living, as well as education. Cultural organization is still largely a matter of subconscious coordination; we possess little of the knowledge yet which would enable us to consciously design new cultures.

The last steps to world unity at the level of social organization may well concur with major fluctuations in the form of widespread upheaval. A design for world unity is clearly outside our present organizational reach. I believe that the decisive impulse has yet to come from the level of cultural organization. It is futile to speculate in which form such an impulse may come—in the form of a charismatic leader, or new Christ, or in the form of a sudden spreading of serenity and enlightenment which would act like oil on the high-going social waves, or even in the form of extraterrestrial intervention (a favorite theme of science fiction). But in one form or the other the final mutation to the global social system is bound to come soon.

If these curves—speculative at least in their future extension—are taken seriously, human life acquires indeed a profound meaning in the perspective of evolution. But from these curves we may also gain a new notion of what noogenesis and involution may imply, terms that had only a very general meaning when they were introduced in Chapter 3. *Involution* may now be understood as the conscious internalization of organization through human design; thus, it is defined at various levels of organization and characterizes phases in which the next higher level becomes externalized, more competitive. Teilhard de Chardin obviously meant the involution of social organization which is in the focus at present. But with social involution, we are moving onward to cultural evolution, and so forth. *Noogenesis,* then, is indeed not a one-level process but in itself a mutatory process and the ultimate manifestation of the principle of "order through fluctuation," as I have already stated.

For the near future, the following basic guidelines for policy design may be learned from Figure 19:

—Cultural organization (syntony), may be helped in an evolutionary sense by furthering cultural differentiation, a pluralism of as many ideas, life styles, and world views as possible. The invention and introduction of new forms of cultural organization ought to become increasingly a matter of conscious design, although the overriding principle for some time to come will be experimentation and testing in a turbulent environment.

—Social organization (relations) is on its way to becoming increasingly a matter of conscious design, and less of competition. This does not mean that dictatorial forms will dominate. Rather, social organization will become the substrate for cultural experimentation and no longer the level at which the battle of human evolution is fought. As people become more interested in cultural organization, trying to understand the guiding purpose of human life, they will see social organization as a manifestation of principles evolving at the cultural level, and no longer as an end in itself.

—Ecological organization (survival) is already largely subject to human design and should remain so to the extent that social organization becomes an ever more conscious affair.

This sounds like a plausible general prescription for the decades of turbulence ahead, a turbulence which will be felt at all levels, even the ecological one. One may even speculate whether social planning might not be used to deliberately introduce fluctuations, and thus to precipitate the mutation to a new dynamic regime under conditions which seem better manageable.[1] Thus, to take a hypothetical example, cities ridden with air pollution may find it easier to introduce adequate legislation if, on a bad inversion and smog day, a large number of army trucks is brought into the city to worsen the situation and to break through a threshold in awareness of the citizens and city fathers. But, I fear, such action would have as little effect in the direction of anticipatory action as do apocalyptic forecasts. In such a situation, real crises of a serious, but not-yet-catastrophic character may have the ultimately desirable effect of facilitating the inevitable mutation to a new regime.

As this chapter is written, late in 1973, the reduction in Arab oil production and the massive price increases for crude oil seem to have quite an effect on the Western world in various ways. Suddenly, long-range thinking with a serious purpose appears as mandatory; the Common Market is tested for the reality of its avowed purposes; and, above all, a long-range look at the evolving global energy picture —available in many forecasts for many years—is now enforced and action stimulated toward a more balanced exploitation of energy sources and the development of alternatives. It is significant that, in this situation, the first politician to emphasize the utilization of the remaining oil reserves for upgrading purposes (petrochemistry) and to call downgrading (combustion) a sin, is the Shah of Iran, a major oil-exporting country. Politicians in Western countries are still waiting for the moment when they will be able to signal their voters the continuation of the old politics of ruthless economic expansion and short-range maximization of resource utilization. But things will never be the same—even if, as I expect, the possible imminent collapse of the old regime will be postponed. The Western world has suddenly and, for most decision-makers and planners, early and unexpectedly run into the first serious limit to its further growth— and even traditional economists who criticized the "Limits to Growth" study of the Club of Rome (Meadows, et al., 1972) on the basis of their belief in a self-regulating market mechanism, may now learn a more realistic outlook. Economy is no longer a matter

1. I am not advocating here a Marxist prescription for "triggering," by means of a revolution, a deterministic movement toward a fixed endpoint. On the contrary, I am speculating here on the possibility of giving life a new chance in suffocating human systems.

of market mechanisms only, but of overall social regulation at the level of the world system. With drastically increased oil prices, the international trade system will now much earlier run out of its possibilities to cope with balance of payments problems. What had been expected, on the basis of the old oil prices, to become an unsolvable problem by 1980, is already one in 1974—in a long-range perspective, this may turn out not to be the work of the devil, but a "kamikaze," a divine wind of change, as which it is partly recognized already in the public discussion developing in particularly vulnerable Japan.

In the period of turbulence ahead, the time scale will probably become very compressed. Evolutionary time in the human domain has already become highly accelerated in comparison with, say, the time of the solar system, or of biological evolution. Figure 6 also suggests that shifting the focus from social to spiritual systems, or cultural organization, reduces stability and increases flexibility. All indications and hunches seem to point to the possibility that cultural organization may pass through its most competitive phase and become subject to conscious design perhaps within a few decades. I believe that this development will depend to a large extent on how soon the fourth level of organization, psychic organization, will make its appearance on the human stage—the basic potential of man to learn how to use his psychic capabilities and powers, which may indeed open up new horizons, and become the principal "pulling force" for cultural design. But it appears futile at present to speculate whether such a psychic organization will then mark the end of the mutatory process and will result in a stable and durable organizational structure,[1] or whether this, in turn, would mean the end of mankind —or even of evolution.

---

1. Insects constitute an astonishingly stable element in evolution on earth. Their phylogeny has not evolved in any significant way for a hundred million years. One may speculate that this extraordinary stability has something to do with a psychic organization keeping insect societies "in place" at an early stage of social organization. Recent empirical research seems to point to such psychic factors. Information about feeding places seems to reach insects much faster and more reliably than through any communication process of reporting, signaling, smell diffusion, or chance sighting. The brothers McKenna (1972) describe how they became aware of messages crisscrossing their consciousness, and which seemed to belong to the insect world, when they were in the tropical forest of the Upper Amazon and in a shamanic state of consciousness.

The evoluionary waves of organization penetrating the human
world may also be viewed as passing through the three levels of perception or inquiry. In Figure 20 I attempted to plot these waves in

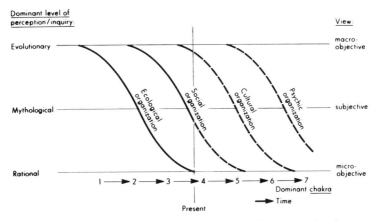

Figure 20. Another view of the waves of organization in
the evolution of mankind (see Figure 19). They may be
interpreted as passing through the three levels of human
perception and understanding. It is at the mythological
level where we feel them most acutely.

such a way that each of them starts at the evolutionary level, with
the realization of an objective, subconscious order, which is an image
of the order of the universe expressing itself through man (again
the law of correspondence). From there, the organizational waves
pass to the mythological level, where subjective qualities and personal relationships come to the foreground as well as the full thrust
of love and human creativity. Finally, each wave moves into the
rational domain of deterministic cause/effect relationships and manip-
ulability. This does not mean that the organization of the human
world will end up in a clockwork society. As waves of organization
glide down into human consciousness and rationality, the focus moves
on to new waves which pass through the mythological level and are
themselves guided by yet unconscious further waves.

The movement from instinctive to reflective modes of organization, that characterizes each wave, may be seen as a genuine evolutionary principle governing the human world, a principle giving
expression to a higher *telos.* Kleist, in his essay "Über das Marionettentheater" (On the Puppet Theater), which I mentioned in Chapter

11, observes that reflection and gracefulness are complementary principles in the organic domain; the splendor of gracefulness shines the
brighter the less reflection dominates. But he adds the deep thought
that in the human domain gracefulness returns when cognition has
gone through "an infinite"—just as the image in a concave mirror,
after having vanished into infinity, suddenly returns close before our
eyes. Thus, says Kleist, gracefulness is most purely expressed through
a human body which holds no self-reflective consciousness at all—
or infinite self-reflective consciousness. "Consequently, I said a bit dis-
tracted, we would have to eat again from the tree of knowledge in
order to return to the state of innocence? Indeed, he answered, this
will be the last chapter of the history of the world." Never has the
idea of design *for* evolution been expressed more clearly.

But we understand now also why the evolutionary process is experienced by man to such an extent as a process full of suffering,
insecurity, and frustration. Since we relate to the world basically
at the mythological level, what we feel above all, is the externalized
struggle of competition in organization. What, through the ages,
man has experienced as the human condition, was this struggle, its
misery and the predicament it implied—first in ecological organization (scarcity), then in social organization (oppression). "Till now
man has been up against nature; from now on he will be up against
his own nature" (Gabor, 1963). The fluctuations of external competition are felt as random, and we endeavor in vain to stabilize
them. It is only when we develop a sense for the evolutionary level
that we vaguely start to comprehend some aspects of the intrinsic
meaning in the evolving pattern of organization, with the "leading"
wave always being beyond our rational and emotional reach.

In the present phase of mankind's evolution, organization is fully
orchestrated over all three levels of understanding, and appears in
its three aspects of cultural, social, and ecological organization. They
are aspects of the one overall organization of the human world and as
such inseparable. It is indeed becoming evident that ecology cannot
be organized separately from social structure, and that both cannot
be organized separately from the systems of values, standards, and
criteria which make up the core of our cultures.

Following the course of the organizational waves in Figure 20, we
gain the following picture: Ecological organization moved from the
kind of objectivity which is built into the basic instinct and ensures
survival, through subjective mythological relations to the fairly
predictable, objective, leverage of technological power. Social or-

ganization moved from an instinctual feeling of belonging and group functions to relations built upon a recognition of the "Thou" in fellow human beings, relations characterized by power as well as by personal love in all its warmth and splendor—and is perhaps moving onward to complex patterns of increasingly conscious role playing, based on pragmatic, impersonal interests and attitudes. We have "nothing personal" against competitors or political enemies. Cultural organization is still in its infancy, following an elusive, but powerful *Zeitgeist* which seems no more predictable than a capricious wind blowing dry leaves in this or that direction, or in circles. If the paradigm chosen for Figure 20 holds at all, it becomes clear that human creativity and love in the time to come will have to express themselves primarily in cultural organization. In the following chapter I shall attempt to sketch a few ideas of what this might imply and point to the vast consequences that such a shift in focus on specifically human capabilities seems to entail.

Here, I should merely like to point out that strong, interpersonal love as many generations have known it and artists of all epochs have glorified it, seems to be on its way out as a general bond in social relations. Rampant individualism in a technological age marked the beginning of such a shift. The increasing separation of love and sex —the suggestive emphasis which the industrial society places on "making love"—is but one aspect of it. As I interact with young people in California, the Western world's foremost cultural and social laboratory, I am becoming aware of a spreading softness, or weakness, or even infantilism—an emerging "cop-out society" which shies away from the strong bonds, the responsibilities, the depths of involvement, and the suffering which human emotions seem to hold when they fully enter into the play. Many of the spiritual schools and centers which attract the young seem to decree the spiritual and mental virtue of nonattachment also as a social virtue; some even preach what they call the "law of consideration"—namely, to do your own thing without consideration of what other people may expect from you. This amounts to a total negation of the role playing mechanism and its basis of self- and mutual expectations which have so far been weaving the social fabric. The implicit ideal seems to come close to a Samoan type of society as Margaret Mead describes it: a light game of pleasure and detachment, an equilibrium world which excludes excellence, engagement, and strong emotions. The indiscriminate use of techniques emanating from the pleasure-and-well-functioning-oriented humanistic psychology and human potential movement (see Chapter 9) seems to result partly in the weakening of human creativity and love in social relations, including the capability of accepting love. For the time being, an overemphasis on vertical individual growth (the awareness of higher levels of consciousness) seems to weaken horizontal social interraction—an imbalance which seems to turn into a threat to the cohesion of human systems, but may also constitute an accelerating factor for mutations to come.

The question then arises whether such a turn toward "becoming like children" may not be viewed as an evolutionary prerequisite for switching human creativity and love from the wave of social organization on to the wave of cultural organization. Is growing sexual impotence a meaningful phenomenon in evolutionary terms? Is our conditioning by hard technology around us, and "human technology" working inside us, preparing new generations for new evolutionary tasks—or are they driving mankind out of its tao? Are these meaningful questions at all? Certainly, in Kleist's terms, gracefulness in evolutionary movement is our best bet in a time of difficult transition. But what kind of gracefulness? The childlike, instinctive kind of gracefulness which may be approached by the elimination of self-reflective consciousness? Are they really opposites? Should we proceed backward or forward—and is there at all a way leading back? With Kleist, I believe in the way forward, in design, as the evolutionary path laid out for man.

What we are witnessing today with regret or indifference or zeal, is a cooler climate in interpersonal relations of all kind, a lack of love which, ironically, seems to become most painfully obvious in those communes or tribal organizations which have gone farthest in developing a new sense of spirituality. Social relations are becoming a pragmatic affair. But the often evoked "universal love" cannot yet be found in other than the most thinned-out and superficial versions. In accordance with Figure 20 I should believe that cultural organization is meant to move from the cold light, the objective idea of such a "universal love" at the evolutionary level, to the warm fire of human emotions and qualities which characterize the subjective mythological level. While the relations of work and everyday life are on their way to becoming pragmatic and cool routine, the relations of human imagination and of the spirit seem bound to move into the furnace of human emotions and of love. It is the time of transition in which the overall temperature seems to approach freezing conditions. But, certainly, what has been hinted at in several earlier chapters, seems to gain new support through considerations of such sweeping nature as proposed here: namely, that the present phase is not just characterized by a transition from one culture to another, but by an evolutionary caesura marking—or at least requiring—a shift to higher levels of human consciousness, understanding, organization, creativity, and love.

Looking at the waves of organization in their present configuration, we understand now also why the change-processes which dominate the physical, social, and spiritual areas differ so vastly from each other. Whereas a change-process "from outside" (locomotion) makes its biggest impact in the physical domain, which is transformed by technology, an "inside" change-process (role playing) appears to be the most effective in the social realm, and a catalytic (resonance, syntony) process seems to govern the spiritual domain. These processes may at present also be understood as characteristic aspects of an overall change-process, composed of ecological, social, and cultural modes of organization. But basically, locomotion is the expression of change at the rational level of perception, role playing the expression of change at the mythological level, and autocatalysis the expression of change at the evolutionary level. As social organization will move on toward the rational level, imposed change will become more important, but then social organization will no longer be crucial for living out freedom. The type of freedom which we associate with the notion of role playing within wide margins, and which is the basic form of relating at the mythological level, will then be searched and lived in the processes of cultural organization.

The assumption of internal factors of evolution in human systems is essential to live out, not only possess, freedom. Gabor (1969) has mathematically treated the general case of design for the future in such a way that the maximum measure of freedom to act is ensured for the following generation of planners. His calculations, for a system under equilibrium conditions, lead to the obvious, if paradoxical, conclusion: Once we have attained and possess maximum freedom, every use we make of it can only decrease it. A ball in precarious balance at the top of a hill, will tend to run all the way downhill if tipped off. But nonequilibrium conditions open up a quite different prospect: The more freedom a system is able to live out and the more freely it can absorb or generate instabilities, the less random its behavior will ultimately be and the greater the chances for a new dynamic order. Human systems are generally of a highly nonequilibrium nature.

The full scope of policy design may now be viewed as it appears in Figure 21, which is an adaptation from Figure 3 (see p. 52) which

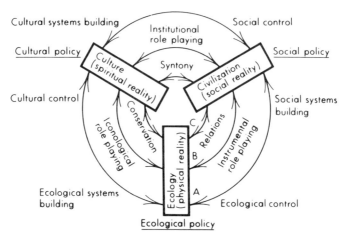

Figure 21. The process structure leading to policy design in the physical, social, and spiritual domains. The outermost circle A refers to the rational level of inquiry, the middle circle B to the mythological level, and the inner circle C to the evolutionary level. The interplay between organization and regulation which leads to policy design is characterized by the alternate clockwise and counterclockwise directions.

had served to discuss human systems. It holds for the whole spectrum from individual to all-embracing societal systems. The two directions of the feedback link leading to policy design—organization and regulation—appear here as a counter-clockwise and a clockwise direction of rotation, respectively. At the interference planes of these processes, policies evolve for the three domains—ecological policy for physical reality; social policy for social systems which, in their interaction, form civilization or social reality; and cultural policy for the spiritual reality.

The three circles of processes in Figure 21 stand for the three levels of perception or inquiry. Most, or almost all, of current policy design is geared to processes at the rational level where organization means the organization (in terms of construction) and building of systems structures, and regulation means the exposure of these structures to control measures of the "bilateral" cause/effect type. Both feedback directions representing policy design at this level are seen as impartial, objective processes amenable to scientific treatment on the basis of empirical (behavioral) evidence and by logical discourse. This is the level at which a fairly good grasp of meaningful ecological policy should still be possible.

But also social policy, designed by processes at the mythological level only, is likely to be considerably off pitch. Here role playing, the way of personal involvement and mutual change, dominates. In Chapter 12 I gave examples for instrumental role playing (by dedicated individuals, groups, or organizations) and for institutional role playing. It is in instrumental role playing that human aspirations enter most powerfully in the pursuit of subjective goals. Values become expressed most effectively through institutional role playing which, in turn, accentuates and accelerates the dynamics in the interplay of different, and often conflicting, values. Iconological role playing, finally, becomes most important in that aspect which may be expressed in terms of the interplay of Jungian archetypes. It becomes increasingly sterile as a culture rigidifies to specific structures—as all cultures tend to—and is at present, in the Western world, dominated by technological icons. Its revival is essential for cultural renewal.

Laszlo (1973 b) supports his confidence in man's capability to deal with regulation at the level of the world system by emphasizing signal/response mechanisms to effect changes in role playing. The signals would be primarily of two types: warning signals of the sort the Club of Rome sounded (Meadows et al., 1972); and suggestions of the types produced by prospective long-range forecasts. Mead (1973) points out the dramatic change encountered when perception, in this way, moves from the rational to the mythological level. Whereas the basic Forrester model (Forrester, 1971) deals with ahuman closed system pressures which define the boundaries set to mankind, role playing in response to signals of the above-mentioned type turns mankind into an open system, open to its own conception of new dynamic regimes. If the first view gives rise to a pessimistic outlook, an open systems view is basically optimistic.

But role playing is gradually losing its creative importance, which it has exercised primarily in social organization. As Vickers (1973 b) points out, role playing works on the basis of match and mismatch signals. Whereas the theories of the 1950s emphasized tension release by innate responses, the 1960s introduced the theory of information, communication, and control of social processes and dealt with innate responses modified by learning. In the 1970s we are in the process of learning to take seriously those responses which are no longer innate, but emerge from tuning in to general evolutionary forces. Syntony, the form of communication established through such resonance phenomena at the psychic and spiritual level, is on the verge of becoming more conscious—after having been effective in a subconscious way in the passing waves of cultures which have united parts of mankind over certain periods of time. In early phases of cultures, e.g., at the Council of Nicaea, people expressed themselves only partially, but understood each other well and in a holistic way. The same happens today between young people from different regions of the world. Young Americans assure me that they are "understood" and that their values are shared in Europe by young people there, even if there is no common language to communicate verbally. Syntony seems to exhibit the same characteristics of holographic communication that seem to govern genetic information and brain functions. This should not come as a surprise if a general holographic theory of mind functions has some basis (see Chapter 9).

As we have learned (though not too well) to design social roles, we shall have to learn now to design systems of syntony. In other words, moving from ecological through social to cultural organization also implies a shift of focus from the rational level through the mythological to the evolutionary level of inquiry. But for this it becomes necessary that we also learn to trust whatever may become visible or felt as a still-unconscious psychic form of organization.

In the following and last chapter of the book I shall attempt to formulate a few thoughts, elaborating on what I have said in this chapter, geared more specifically to the "coordinative conditions" in the physical, social, and spiritual domain. This last chapter is meant to put forward attitudes and principles for policy design which cannot yet be perceived with any clarity and precision, not to speak of their generation of an explicit methodology. But I hope nevertheless that it will sum up and illuminate in a somewhat more practical light what I have tried to trace in this book.

(15)

CENTERING REALITY

A look at Figure 21 might suggest that we have a sufficiently supple and consistent design mechanism at our disposal which would permit us to neatly balance out policies for all three aspects of reality—physical, social, and spiritual—and to bring them into consonance with each other. But we know only too well that this is not what our current attempts at policy design reveal.

Why should it be so difficult to center reality once we have learned to bring all the processes in Figure 21 to bear upon policy design? One reason certainly is that we have not yet learned to bring all of them duly into play. Our overly scientific approaches to matters of policy, of designing our lives under viable preponderant themes, tend to restrict us to the rational level of inquiry. In Chapters 12 and 13 I tried to sketch potential modes of breaking out of the deterministic cause/effect straitjacket by means of the systems approach and of transexperiential inquiry. And in Chapter 9, I surveyed some of the approaches toward a better grasp of the processes at the evolutionary level.

But there is also another important reason for our failure to balance human reality. It concerns our own clumsiness in designing the processes themselves which enter into policy design. In the direction of organization, we mainly introduce processes which are characterized by positive feedback, or "feedforward"; it is perhaps more suggestive to use here Vickers's (1970) term "self-exciting process." Many processes of organization form a self-exciting system which, so to speak, spirals up and away once it has been triggered. Our economic system is geared to continuous exponential growth under the label of "Keynesian economics." The "car culture" in America (with Europe apparently soon to follow) has reached a stage where the widespread use of cars has led to the whole service structure of society becoming geared primarily to serving cars, not people—with the result that it becomes increasingly difficult to live without a car which, in turn, leads to a more ubiquitous role for the car, and so forth. The population explosion and the undernourishment of an increasing number of people in the world lead to frantic attempts toward increased food production by better use of arable land and by new technology (e.g., the "green revolution," exemplified by "miracle rice" species) which, in turn, removes the principal limitation from population growth and results in ever larger numbers of people living at the fringe of starvation, and so forth.

In the direction of regulation, on the other hand, we know only how to apply negative feedback, how to bring dynamic developments back to preset, fixed, or gliding reference values. The ideal here is that of homeostasis, of peace with an environment which we thus let govern our lives. If the economy gets heated up, governments apply their instrumentarium of monetary and fiscal measures, in the —increasingly futile—hope that the mechanisms described by economists will lead to the desired overall negative feedback control.

But what is so fatal about our use of these two basic modes of influencing complex dynamic developments is the dualism built into their interaction as we perceive and conceive it in our minds, models, attitudes, and institutions. We act as if we were rowing a boat with one oar at a time, putting all our energy into a succession of onesided strokes, always overcorrecting deviations from our general course and vaguely and slowly zigzagging forward. This removes us further and further from the ideal of self-regulation—of designing and operating a system in which positive and negative feedback processes would counteract each other sufficiently to warrant only a minimum of human intervention.

So-called "natural ecology"—the artificial image of life on earth in the absence of man—acts as a self-regulatory system. Even human societies at the stage of simple technology were self-regulatory with regard to filling their physical and social space. Limitations in food production and high infant mortality saw to that. The only "positive feedback" way out of such a situation was to acquire new land— by dangerous migrations, such as those to new archipelagos in the Pacific ocean, or simply by waging wars against neighboring land-users. Western ethics together with Western food and health technology, of course, destroyed this self-regulatory idyll, which may not look so peaceful and desirable after all. Nature is not sentimental, but humans are excessively so—nature is interested in species and evolution, man in individual life.

But our more recently formed attitude toward what we call "the environment"—a notion presumably standing for something like the earth not counting ourselves—shows drastically our dualistic approach to design and the nebulosity of the myth of self-regulation, as we hope to make it work. We accept responsibility for regulating the environment, but we are hesitant to include ourselves in the system to be regulated. Or we do so in the half-hearted ways which characterize our current approaches to social policy.

Holling and Goldberg (1971) and Holling (1973) have pointed out that in the initial phases of any kind of ecological organization (e.g., among the species appearing in a large area stripped of vegetation) the gradual increase in the variety of species and in the complexity of interaction is accompanied by an increase in the resilience of the system and a decrease in productivity. Under stable conditions a self-regulating ecosystem evolves in which species control each other (e.g., birds and insects—if the insect population soars, the bird population increases until it has brought down the insect population; this, in turn, leaves food only for a smaller number of birds, and so forth). There are upper and lower boundaries for the domain of

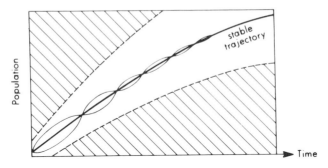

Figure 22. Example of a bounded stable trajectory as it is found in an ecological succession (adapted from Holling and Goldberg (1971)).

stability and the regulatory forces tend to be weak until the system approaches a boundary. Such ecosystems are not efficient in an optimizing sense "because the price paid for efficiency is a decreased resilience and a high probability of extinction" (Holling and Goldberg, 1971). Figure 22 shows the example of an ecological succession (an evolving ecosystem).

Man's interaction with nature, from primitive agriculture to industrial activities, is characterized by the top criteria of efficiency and productivity which high technology makes possible to apply.

The result is the contraction of the boundaries of stability. Trying to maintain an equilibrium by more and more efficient means of control becomes increasingly precarious. Any excursion of a systems factor coming close to a boundary, risks piercing this boundary by virtue of the impetus imparted by high efficiency, as well as throwing the entire system into a disintegrating mode. Hard and soft technology, and the institutions of society applying it toward the goal of high efficiency (e.g., business) at the same time narrow down the boundaries and increase the danger of piercing them. Within such narrow boundaries a negative feedback loop may quickly flip over to a positive feedback characteristic, and vice versa, without our even becoming aware of it before its grave consequences appear in the overall systems behavior.

So far, when we speak of piercing boundaries, we mean the boundaries of ecological organization. Piercing them may not result in the disintegration of our ecological systems but constitute just another aspect of a mutation from ecological to social organization. Human systems develop then within new boundaries which are no more rigid but become redefined with every mutation toward a more complex social regime. But what has to be recognized now is our integral participation as regulators in the systems to be regulated. The ecological notion of self-regulation—no or little human intervention—is devoid of meaning here. Self-regulation in the realm of social organization inherently implies our moral engagement in regulating human systems which include ourselves.

But within the new and expanding boundaries of social organization, regulation becomes dramatically accentuated by the acceleration in many factors of physical and social development, with doubling times in many cases down to ten years and less. The kind of regulation required may be compared to that applied to a nuclear reactor. It is no longer sufficient to react when a development has reached its alarm level—it has become necessary to act in an anticipatory way. If a fuel rod in a nuclear reactor heats up, or if the neutron flux starts to rise, control rods have to be shot in instantly, when the indicators are still far from the alarm level, in order to avoid "overshooting" of the latter. Of course, there are other built-in modes of bringing negative feedback into effect—such as the explosion and subsequent subcriticality of the reactor. Analogous kinds of equilibria also are in the offing for human society if regulation by anticipatory action fails to become effective in time. In some areas it would seem that such a failure is already a fact and that the consequences in terms of major crises are approaching. Anticipatory action, and planning conducive to it—as opposed to "muddling through"—seem to run against our present cultural preoccupations, as Michael (1968) has discussed in a penetrating way.

It seems that we have also pierced the boundaries of social organization and act now within the boundaries of cultural organization (as I also pointed out in the last chapter). But instead of integrating physical and social planning with cultural design, we put our faith in what may be termed the "technological forward escape." For every nuisance created by technology we propose a countertechnology—the "endless technology/countertechnology chain" which Dubos (1969) has denounced. We attempt to drive out the dangers of efficiency by more efficiency, the devil by Beelzebub, making human life on this planet ecologically more and more precarious. The alternative at the ecological level, as Holling and Goldberg (1971) point out, would be the replacement of an equilibrium-centered point of view by a boundary-oriented view as a conceptual framework for man's intervention in ecological systems, be they natural or man-made. Such a framework "changes the emphasis from maximizing the probability of success to minimizing the chance of disaster. It shifts the concentration from the forces that lead to convergence on equilibrium, to the forces that lead to divergence from a boundary. It shifts our interest from increased efficiency to the need for resilience." One might add, it also provides increased chances for maintaining sufficient nonequilibrium to bring the principle of "order through fluctuation" into play.

In the area of social policy we seem to use heavier and heavier oars, so that we have to use both hands to handle them alternately, one at a time, giving our boat more and more "efficient" pushes to the right and to the left in order to offset the change in direction the last stroke has imparted to the boat—ever more losing balance and control over our course. The positive feedback "strokes" dominate in the name of progress and have thrown the world into a vortex of growth in all areas. The automatism of a self-regulatory balance between positive and negative feedback loops does not work anymore in an era of heavy technological intervention.[1] How we handle

1. Naive beliefs in "natural" combinations of positive and negative feedback loops are often voiced in this desperate situation. Some demographers, for example, advance the idea that economic development, which would bring the

our two oars is really too awkward. It corresponds to a particular answer to that basic question which I have evoked in the first chapter of this book: Can we assume that freedom, which we cherish as our highest social good, is something *given*, something whose existence has to be merely guarded and protected effectively in order to be preserved? But is freedom not also, to use Vickers's words (1970), a "social artifact," something which has to be actively designed and built in order to come into existence?

One answer, which is the answer implicit in most of today's management literature, holds the former view. It is based on the "free enterprise" ideal and where it becomes obvious that such an ideal might lead to disaster, it proposes something like a "management by exception" game, the creation of a "free space" for business to play out its activities as it sees fit, whereas the rest of the space is excepted from the game. Presumably, the rest of the space is taboo or reserved for society minus business, so that we have two separate systems interacting with each other by exchange only—again a manifestation of the fatal dualism in the Western system of thought and action. Here it appears in the form of a separation of producer and consumer activities.

Who, then, carves out the free space; in other words, who designs the total system of society including business? That is the role of government, the view holds, the role of the only institution which has a design mandate for society. But this setup does not work well as we see everywhere nowadays. And we begin faintly to understand why it does not work well. In many areas, especially in areas having to do with technological innovation and its introduction into the systems of human life, government is set up primarily to regulate by *reacting* to the actions of the private sector, not so much by acting itself. It tries to maintain some dynamic stability in the overall system of society by attempting, usually in vain, some homeostasis

---

whole world to the current highest economic levels in industrialized countries, would automatically lead to a stabilization or even decline in population figures —see the negative balance in wealthy West Germany. The trouble with this beautiful "natural" mechanism is not only that it superficially correlates lower birth rate with being well-off (there are more profound cultural and social factors at work in predominantly rural societies), but that it also would need time and very wide boundaries to be acted out. There is no way of getting the world with its rapidly increasing population over such a global economic "peak" without wrecking it in the process. Such a model is no more realistic than the proposition of cannibalism to solve simultaneously the problems of food production and population growth.

with a changing environment. If corporate power becomes of concern, there is an antitrust law; if an alleged "technological gap"
becomes of concern, incentives for industrial mergers are created;
if exports sag below imports, a monetary devaluation is considered;
if pollution becomes of concern, there is an antipollution law; if
people get concerned about technology more generally, the idea of
"technology assessment" comes to the foreground of discussions.

All these types of *post festum* reactions belong to negative feedback control, bringing variables back to some given reference value
—if it is powerful enough to do so.[1] But who makes them deviate
from these reference values in the first place? Who exercises positive
feedback in our type of society? Who unsettles things, innovates,
pushes in certain directions and holds back in others? It is, of course,
mainly the private sector which is so inventive in using the "free
space" it is granted; and control usually comes too late and is becoming increasingly ineffective. The private sector is, in general, not
overstepping its boundaries; it pays taxes and abides by the law; it
sets up foundations and contributes to charities; it shifts from the
"rule of the wealthy families" to widespread and anonymous public
ownership—and yet its self-excitation and the dynamics it imparts
to society give rise to serious concern, which is even shared by quite
a few among the more enlightened business leaders, who are themselves sufficiently in tune with the reality they shape.

What I am describing here is a mismatch in institutional roles—
a mismatch in the roles assumed today by the institutions of business
and of government, to be more precise. Whereas government is
mainly set up to regulate society as an *adaptive system*, that is, as a
system adapting to changes in the environment in order to survive
(see Chapter 4), business acts, so-to-speak, outside society (though
interacting with it, of course) as an *inventive system*. Business largely
sets its own goals and policies, exercises much leverage on the environment, and is capable and permitted to actively shape the environment. The environment it shapes includes the rest of society,
the subject of public regulation. Governmental regulation deals
only incidentally with the shaping of the environment. But in this
situation, an integral design approach may come to be based on a

1. According to a study for the U.S. Congress (1973), 278 billion dollars
were freely available in private hands at the end of 1971 (mostly with multi-
national corporations), i.e., more than double the total currency reserves of all
Western countries together. The positive feedback making itself felt in the
movements of private money in times of monetary crises, can hardly be
countered anymore by the government reserves meant to reestablish balance.

new type of politically responsible leader, the "public executive" (Cleveland, 1972) who is a top manager of business or a university or a foundation and understands his task as one of close interaction with, and responsibility for, society at large.

Whereas man's ecological organization is now principally affected by conscious internal factors, by planning and purposeful action from agriculture all the way to the highly technological urban milieu, social policy in the Western world is still formulated in line with predominantly Darwinian principles. "Free enterprise" is only the most widely known among them, with wars between nations being the most atrocious one. Competition is fostered as an indisputable and necessary ingredient of a liberal social mechanism and protected and even imposed by antitrust legislation. Coordination is still officially suspect. The peculiar mechanism which social Darwinism employs in beating a reluctant retreat may be compared to the building of fences around a piece of land containing rampant vegetation. As social interests seem affected, the fence is closed in more and more. But no redefinition of the area allotted to nature's "free enterprise" will make a garden out of that piece of land and the ecological competition allowed to continue on it will hardly produce many of the products which coordination of growth, planting, weeding out, and cultivating may expect to yield. If we want a rose, we have to plant one—in ecological as well as in social organization.

The Darwinian credo even seems embedded in recent ideas for technology assessment. As if technology, totally a man-made artifact, would only grow "wild." As a matter of fact, practically all technology has developed around normative or at least teleological ideas (see Jantsch, 1967), in other words through coordination of various inputs and subtechnologies toward a goal. Every technology is "planted"—yet the concept of designing a "garden" in which technological growth might be coordinated, is still one at which every science policy-maker throws up his hands. The idea of technology assessment, like "free enterprise," is still elaborated primarily with a flexible fence around wild growth in mind. Of course, behind this failure, or despair to coordinate, lies not just a misunderstanding of the nature of the man-made physical milieu, but also the failure to coordinate the people which "plant" individual technologies.

If we regard—and why should we not?—business, government, and all the rest of society as forming one system, we may say that currently there is always a subsystem in the whole system designing and organizing its proper subsystem activities, whereas another subsystem tries from its side to maintain a precarious balance for the overall system. This would not in itself necessarily lead to the overall system's deterioration if the acting subsystem would try to act in the interest of the whole system, as the regulating system supposedly tries to regulate in the interest of the whole system. In oher words, the central question is, can the institution of business—can all institutions of society—play roles which are in the interest of society as a whole? Can they all play roles *within* the system of society? The systems approach, as I outlined it in Chapter 12, constitutes an attempt to view roles of individual instrumentalities and institutions in such a societal systems context.

Taylor (1973) proposes an interesting geopolitical evolution model, which may be understood as a substructure of dynamic regimes, forming what I called the social mode of organization. This model correlates specific levels of societal and political organization with characteristic stages of environmental control expressed in terms of science, technology, transportation, and communication. The same levels also correspond to the ascent of system scope from tribal to global systems. Each new stage is first characterized by positive feedback processes—organizing its dynamic key elements—and subsequently, as constraints are encountered, increasingly by negative feedback. I believe that this dynamic sociopolitical model is indeed capable of describing qualitative mutations within the social mode of organization, if it is understood that several levels are in play simultaneously. As a higher geopolitical level is dominated by positive feedback establishing a new dynamic regime, lower levels become coordinated by negative feedback. As national systems become more and more subject to homeostatic control in the presence of stringent constraints, regional systems and the emerging global system reach out with positive feedback, groping for their viable structure. Another point of view shows instrumentalities becoming subjected to negative feedback control (e.g., the curbing of "free enterprise"), whereas institutions are changing and organizing toward new roles.

I have written elsewhere (Jantsch, 1972 a) about the mismatch in societal institutions managing science and technology, and elaborated some ideas for an integral responsibility vis-à-vis society to be shared by all the institutions primarily involved, i.e., by government, business, and higher education. The essence of my proposals, in terms of the mythological level, is to row "with both oars," positive and negative feedback at the same time, and to design patterns of institutional interaction in which every participating institution would share in the undivided, nondualistic responsibility for the whole system—business as outlined in Chapter 12; and the university by focusing on "education for the self-renewal of society" and embracing education, research, and service to society as integral aspects of learning.

Here, I should like to comment briefly on institutional role playing in the political domain. If spiritual norms are not widely shared, political norms, which may be set in arbitrary ways, tend to fill the vacuum. If there is no effective cultural order, no shared appreciative system, political order becomes imposed. In primitive cultures, which are closely tied together by sharing in the same mythical culture, there are no laws, but there are social taboos—which act much more powerfully than any law might.

We are currently witnessing, in Western countries, a process of liberation from imposed political order, in which we may recognize perhaps the beginnings not only of a newly emerging cultural order —a new pearl in the string of cultural orders over the past millennia —but of the idea of a *cultural design*. This liberation becomes visible in the following aspects of political differentiation:

—Increasing rationality, moving from the first step of rationalization of political rule, namely bureaucracy, to the second step of "objective" and "scientific" approaches to political decision-making and the rule of the expert exercised through advice giving (Habermas, 1966). Although in itself perhaps dubious, and even dangerous, this development nevertheless is in line with a more conscious approach to social organization.

—Increasing horizontality of the planning and decision-making processes. Cleveland (1972) has pointed to this shift in emphasis from vertical to horizontal processes, as organization systems become too complex to be handled by the traditional organization pyramid. More generally, the old class struggle between vertical orders of incompatible levels— representing a stratified hierarchical system in which control emanates from the top—has in Western society gradually given way to more horizontal institutional role playing, to a multiechelon system based on the coordination of multiple values and goals. One of the most important consequences is the shift in recognition of government as the representative organization of the "ruling class" to government as an institution of equal rank with other institutions of society.[1]

---

1. As this is written, the Watergate affair in the United States provides dramatic evidence of the crisis of legitimacy and its inflated claims—and also of the failure of the government in power to recognize the writing on the wall.

—Encouragement of dissent and the introduction of instabilities (Cleveland, 1972), as incredible as this may sound to some ears. It is the historical merit of student unrest and protest against wars that the scene has, indeed, changed drastically as compared to only twenty years ago—which period, in the United States, was characterized by an intimidating McCarthyism on the one hand, and "sweeping-under-the-rug" attitudes on the other.

—The demand for a discussion of society's future in terms of values. In this context, I believe that the function of the head of government will increasingly become one of leading a dialogue with the people-at-large in terms of basic values involved.[1] It will no longer be sufficient for governments to justify their basic decisions only in retrospect, and only when election time is approaching. The dialogue (perhaps involving new institutions, not just the traditional mass media) ought to be continuous. On the other hand, Parliament, which has in the past assumed a role of "representing the values of the people"—a claim which cannot be upheld any longer—would fit into such a new scheme as the body of full-time experts ensuring that governmental strategies (e.g., legislative measures) are in conformity with the thrust of the ongoing value dialogue; this is a highly technical job and requires complex management.[2]

What I have outlined as dilemmas in the areas of ecological and social policy, point more and more dramatically to the necessity of integration in a framework of continuous, or at least "batch-wise," cultural renewal. As I have already said in this book, I understand by culture at least a shared appreciative system, plus a communication system through which sharing becomes possible. Culture in-

1. Prime Minister Trudeau of Canada is one of the few contemporary statesmen to whom a political discourse in terms of values comes most natural. When he was Minister of Justice, he once seemed to fight a losing battle in Parliament for the liberalization of sections of the penal code concerning the domain of sexuality—until he exclaimed: "The state has no business in the bedrooms of the nation!" The appeal to basic values led to a change in attitude and Trudeau won his case by a landslide.

2. The numerous proposals for "instant voting" are, of course, technically feasible. But it would not make much sense to turn the process of strategic synthesis and decision-making into a more or less empty routine. "Direct democracy" in Switzerland, where people have to vote on several dozens of issues at the national, cantonal, and community level each year, has resulted in oversaturation, lack of interest and low participation (in many cases as low as only 30 per cent). On the other hand, Gallup polls—a form of instant voting by sample—taken during the flare-ups following the assassination of Martin Luther King in America, first resulted in answers 40 per cent of which may be classified as "fascist"; a few days later that percentage was down to 10 per cent, indicating the susceptibility of instant voting to situation-contingent emotional reactions.

cludes a view of the world, through the prism of values, plus modes of bringing the latter into play effectively, sharing and passing them on. In a static world, culture is translated directly into norms through religious or ideological ethics.[1] But in a dynamic world, we need a set of flexibly applied regulatory principles; in other words, a policy.

As I pointed out in the last chapter, cultures and civilizations hang together as long as social organization is the "leading" form of organization. They cannot be designed separately, as a naive approach to Third World problems seems to hold, taking Western technology to cement an indigenous culture; or just a little bit of Western technology, not enough to topple the indigenous culture; or using Western political systems under the umbrella of an indigenous culture. The example of culturally schizoid Japan which has thrown her own refined culture overboard within a century, is here to learn from it.

So far, social organization tried to secure a stable culture "from below" by means of a religion, an ideology, or simply a set of strong social taboos. But as we move on toward a potential for cultural design we also shift the focus of attention to the evolutionary level, to syntony. It is in the domain of culture, in the spiritual domain, that those instabilities become introduced today which elevate the dissipative structures of man's highly nonequilibrium world to new dynamic regimes.

Increasing social stability can now only be secured "from above," by cultural mobility. To impair the latter, to try to save the cultural and social *status quo*, will be the greatest of all sins; it will have to be paid for by the conditions of Orwell's *1984*, one of the most sensitive, accurate, and still not improbable forecasts ever made. Man as a "crisis provoker" (Calhoun) is a channel for the introduction of cultural instabilities, of the continuous search for cultural renewal. To invent/design/find new anthropomorphic worlds is man's most basic urge—finding them becomes the foremost aspect at the evolutionary level of inquiry. But what makes man so restless in his spiritual space, ever overthrowing the orders established in physical and social space? At the rational level we may readily identify internal and external factors which tend to weaken old cultures and give rise to new ones—pleasure-fixation and hedonism,

---

1. The old Egyptian culture may serve as an example of such a static culture. Evolutionary time runs at a different pace in such a practically history-less culture; the Egyptian culture lasted for five thousand years, longer than cultures with much historical movement.

Baroque expression, corruption, social stress, rigidifying imperialism, internal invasion by new ideas and beliefs, and external invasion by enemies; Toynbee (1972) has described them in a study of more than two dozen cultures. But what appears as a cause from one point of view is but a symptom viewed from another angle. At the mythological level, we find the need to build a livable home in a hostile environment, to establish personal and personified relationships with the forces surrounding us and acting *upon* us. The arts have their origin in this need to turn the whole world into a partner to talk to, to move from soliloquy to dialogue—and so has technology. At the evolutionary level, cultures take shape by expressing the forces acting *within* and *through* us.

The idea of *design*, then, marks a radical departure from the old idea of *control*. Classical control theory is geared to stability and to the optimization of specific parameters under a given set of fixed constraints. The newly emerging idea of hierarchical control consists largely of the "optimization of the level of constraints in forming new structural levels and in the optimization of loss of detail in forming new descriptive levels" (Pattee, 1973). This is another way of recognizing the life of self-realizing and self-balancing processes in evolution. It also sheds new light on the old issue of free will in the context of determinacy of laws. In its essence, it seems to say what evolution keeps expressing and what I have adopted in this book as the core notion for design: Give life a chance.

In mankind's present phase of evolution we may recognize specific forces urging us on toward cultural renewal: At the rational level it may be, as Beer (1971) suspects, the recognition of the "meta-problem" that we are no longer capable of solving our mounting problems. At the mythological level it may be our facing up to the profound moral issues inherent in the dynamic situation—and "letting things go" without intervention is a heavy moral decision, too. "Now I understand what planning is about," one of my students in Berkeley exclaimed when we had discussed the necessity of cultural change for making normative planning possible at all. "It is not about the negotiables but about the nonnegotiables!" At the evolutionary level, finally, we seem to feel the rising "psychic temperature of the earth" of which Teilhard de Chardin already spoke twenty years ago. It expresses itself in increasing syntony among the young, as vague as its contents may still be, and in noticeable shifts of consciousness, of which Reich's (1970) Consciousness I, II, and III are holistic, but perhaps somewhat naive expressions. Mind-at-large seems to be on the move again in a very significant way, and we along with it—there is little concrete to be said about it at this stage. "Wenn ihr's nicht fühlt, ihr werdet's nicht erjagen" (If you don't sense it, you will never grasp it), Goethe warned the overly rational.

Out of the current loneliness at the mythological level, the "chilly" godless and loveless period through which we move, and the intensifying experience of evolutionary forces inside and around us, there may spring a new warmth, a universal love of the sort Giraudoux's Suzanne felt after having been rescued from her South Sea island and brought back to the world of men. On the island, she had not given in to the temptation of building herself a mythological "home" of models and myths (see Chapter 5); she had remained totally fluid, streaming into the intense life around her. Back in the world of people, she found to her own amazement that she loved everyone:

> My solitude had raised the pitch of that vague indifference that we feel for our fellow-men, and it began at love. In all of Lewis Island I tried to find one human being that I could not love. . . . But to contemplate for five minutes the iris of a pearl fisher, which had become microscopic because of diving; the malice in the eyeball of a bishop; the faith in a God more beautiful than the most beautiful Kanaka, in the pupil of a Kanaka, and not to feel oneself transported with love by them! The approach of every human being intoxicated me like a pipe of opium. I restrained myself from kissing them and breathing their breath, their sparkling eyes. I stopped before each human head as before a cage and whistled to birds.

The crucial moment in Suzanne's process of becoming human arrived when she encountered the first man in whom she found nothing to love. The flow of life for the first time selected and accepted form—a total reversal of the usual process of becoming human by filling some love into the empty bottles provided by education and socialization processes. Maybe such a reversed mode of becoming human, by bringing form to hold and express feeling and emotions, will characterize the evolving cultural organization once the stream of specifically human life has broken through the forms of knowledge and behavior.

The message of love, in the sense of all-embracing compassion, is not new; it has been given by the New Testament almost two thousand years ago, and by Buddhism even earlier. But it has not yet fully penetrated the human fabric. It has nourished the rare and beautiful flowers of personal love, celebrated by Stendhal and culminating in his top category of *l'amour passion*. But social organization up to now has been primarily governed by energy received and dissipated at the power chakra. Looking at Figures 19 and 20, I have some confidence, or at least hope, that cultural design, as it moves into the focus of man's activity and toward the subjective level of mythological relations, will be governed by the heart chakra. I foresee also a renewal of the arts, not only as the noblest expression of human emotions and mythological bonds, but also as the strongest integrating force between the objective and the subjective levels, as a reflection of evolutionary wisdom cast in the most profoundly human terms. Like love and human warmth, the arts also seem to be somehow "suspended" in the present chilly transitional period leading mankind onward from social to cultural organization.

In short, I foresee and hope that objective light will initiate, but subjective warmth nurture, the forthcoming design of cultures in increasingly pluralistic forms. What I see emerging, as social space is rapidly filling up, is not the one unified world culture which—after the trauma of two world wars—had been looked for as the possible apotheosis of objective and "value-free" science, Western style. Rather, I see emerging a system of pluralistic cultural processes which may be termed cultural or iconological role playing (see Figure 21). By all that we know of systems, such a richly orchestrated "cultural ecosystem" should then also be better capable of adapting itself. With the focus of cultural organization moving toward the mythological level, role playing becomes the predominant mode of initiating cultural change. Iconological role playing will be tied together by a growing recognition of basic, transcultural humanness, and by the feminine principle, or love in the widest sense, which Teilhard has called the "ciment unitif," the unifying glue of mankind.

Like instrumental and institutional role playing, this new dynamic system for the expression of human life and human roles in evolution, will not operate without risks, dangers, excesses, distortions, and aberrations. But if it is "driven" and coordinated by the following wave of human organization—that dimly visible psychic organization which I suspect to be called upon to grow stronger and back up the more conscious approaches to cultural design—we may rest assured of the continuous *bliss of instability*, of life breaking through any tentative order which we may design and implement to live in but temporarily, of being with the stream of evolution, of *being the stream*:

> Consequently: he who wants to have right without wrong,
> Order without disorder,
> Does not understand the principles
> Of heaven and earth.
> He does not know how
> Things hang together.

Chuang Tzu

Epilogue

God writes straight even on crooked lines.
Paul Claudel

PLANNING AND LOVE

Design and evolution work by the same basic process of mutation toward new, generally more complex, dynamic regimes of
systems. The principle guiding this process is the generation of
conditions permitting the production of entropy and thus fostering
metabolizing activity in the widest sense, the exchange of energy
in all forms in which it becomes manifest—physical and psychic
energy, matter and information, complexity and order, consciousness and mind, emotion and spirit. The basic mechanism underlying
the mutations to new regimes seems to reveal itself in all domains
as "order through fluctuation," Prigogine's great discovery. It has
been found and mathematically formulated for the nonequilibrium
physical world and is being investigated for its role in biological
organization. But in this book I have pointed to it also in connection
with the shaping of human consciousness, the evolution of human
systems,—social as well as cultural—and the overall organization of
the human world.

Most importantly, however, "order through fluctuation" also
seems to underlie the human design process. Both design and evolution work through the same self-realizing, self-centering type of
process. Physical, social, and spiritual reality in this view merge into
one seamless ensemble of process—the process of evolution, of which
human design is an integral aspect.

A static world is based on the notions of equilibrium and of a
dualism between irreconcilable opposites. But an evolutionary world
of processes resolves the dualism between such notions as de-
terminism/free will, objectivity/subjectivity, masculine/feminine,
structure/process. In Chapter 3 I briefly discussed the ambivalent
nature of evolutionary processes which are both deterministic and
finalistic—deterministic viewed in the direction from the event into
the future, and finalistic viewed from a teleological or variational

principle guiding the direction of the process. Deterministic and finalistic principles may also be considered to be at work at different hierarchical levels, with local and short-range determinism subordinated to and coordinated by finalistic policies which regulate the dynamics of the movement. Under a policy (a theme underlying a life), many targets may be pursued in sequence or simultaneously; to reach them, deterministic processes have to be set up at the corresponding level—such as in product development, or construction programs, or any operational short-range plans—but they are nonetheless hierarchically guided, selected, and coordinated under the broader umbrella of a strategy and the regulating principles of a policy.

We may say that in good design evolution concurs with human will—or human will with evolution. What, then, forms the distinction between planning and evolution—just the planner, the human versus the divine? And do they form a pair of competitors, in which one side eventually will win and in which human freedom invariably will draw the shorter end?

In a world of monotonously increasing entropy, human will would be viewed, together with life, as struggling against the stream—as waging a heroic but ultimately meaningless fight. In an evolutionary, self-renewing and self-elevating (though, eventually, self-terminating) world, in contrast, subjective will joins hands with objective will. By *will* I understand here simply any organizing force. "Action," states Roberts (1970), "is not a force from without that acts upon matter. Action is, instead, the inside vitality of the inner universe—it is the dilemma between inner vitality's desire and impetus to completely materialize itself, and its inability to completely do so."

Subjectivity and objectivity become two aspects of the same evolutionary process, a process of the kind which has found its most magnificent expression in ancient Greek tragedy, especially Sophocles. There, man recognizes the laws of the gods, the objective "internal factors" of the process, which may set the stage for a tragic conflict—but he does not suffer them passively, he accepts the process out of his own free will and thereby gains true freedom. Antigone goes to her death not because she is driven but because she has made a free decision. In this act of freedom, subjectivity and objectivity, free will and determinism, fall together. Antigone's subjective will becomes identical with objective will: "To partake

in love, not in hatred, is my share." To bury Polyneikes conflicts
with the laws of the state but conforms with the overruling laws of
the gods—it is commanded by objective will, by universal love.

Planning, the subjective activity of organizing the human world,
is then the principal manifestation of *subjective will*; and *love,* the
attractor in evolutionary organization, the manifestation of *objective
will.* That human love is a manifestation of will is a recurrent theme
in many of the more profound theories of love, from Ortega y
Gasset to Fromm (1956). "For in every act of love and will—and
in the long run they are both present in each genuine act," states
May (1969)—"we mold ourselves and our world simultaneously.
This is what it means to embrace the future." But love does not arise
from subjective will, or not alone from it. It is part of a universal
force whose role in evolution it is to free the spirit from the material
manifestations in which it has got caught:

> Wenn starke Geisteskraft
> Die Elemente
> An sich herangerafft,
> Kein Engel trennte
> Geeinte Zweinatur
> Der innigen beiden;
> Die ewige Liebe nur
> Vermag's zu scheiden.
>
> (When every element
> The mind's higher forces
> Have seized, subdued and blent,
> No Angel divorces
> Twin-natures single grown,
> That inly mate them;
> Eternal Love, alone,
> Can separate them.)
>
> (Goethe, *Faust II*, Closing Scene,
> The More Perfect Angels)

An evolutionary perspective almost reverses the categories of a
rational world (see Figure 14). Where, in a rational, logical world,
the feminine is "groundedness," the abysmal side of human nature,
man's roots sunk deep into the earth, and the masculine is the mobile
side, mind freely roaming on earth, in an evolutionary world the
feminine is instability and process, and the masculine acts out power
through stability and structure. If in a dualistic world mind is

masculine and matter is feminine, in an evolutionary world the unbound spirit uses the feminine as its agent and its material manifestations are masculine.

In Figure 14 (see p. 178) we may now identify the masculine with planning, or subjective will, and the feminine with love, or objective will, in all four dimensions of centering. But, whereas the objective will constrains the free roaming of subjectivity of the earth, it is at the same time the force which attracts us toward higher states of evolutionary existence. Planning aims at fixation, integration, materialization, stability. Love frees the spirit, separates, introduces instability and degrees of freedom. As I have already mentioned in earlier chapters, in a nonequilibrium world—contrary to the classical, equilibrium view—more degrees of freedom result in *less* randomness and more evolutionary order in the process. Planning stands for the masculine, for Apollonic clarity and explicitness, for the focusing and acting out of human energy, whereas love stands for the feminine, for self-abandonment to nature, which keeps the evolutionary flow moving. What planning solves, love dissolves; where planning defines levels on which to place concrete realizations, love is the upward sweep which breaks them up, the event which makes all structure transitory, the true reality in an elusive world of change:

> Alles Vergängliche
> Ist nur ein Gleichnis;
> Das Unzulängliche,
> Hier wird's Ereignis;
> Das Unbeschreibliche,
> Hier ist's getan;
> Das Ewig-Weibliche
> Zieht uns hinan.
>
> (All things transitory
> But as symbols are sent;
> Earth's insufficiency
> Here grows to Event;
> The Indescribable,
> Here it is done;
> The eternal Feminine leadeth us
> Upward and on.)
>
> (Goethe, *Faust II*, Closing Scene,
> Chorus Mysticus)

The evolutionary process alternates between periods of relative stability at a specific level of organization, and periods of increasing instability which lead to stabilization at a higher level. Masculine and feminine periods alternate to ensure the life of the process. Stabilization makes it possible to formulate and organize, to build a culture, to act out power, to force the environment into partnership or subservience. The evolution of humanity followed such a dualistic or sexual pattern of development. But the dualism evoked here is not separating earth from heaven, man from God—it is the unfolding of an innate dualism in the divine principle itself, as Jung (1961) formulated it at the end of his life. It is the very yin/yang principle of all unfolding energy.

But the ultimate freedom in an evolutionary process would only be attained if there were no periodic leveling off. If sufficient instability is always introduced and maintained in the process, it might be possible to achieve a continuous upward sweep. The poetic image of this purest ideal of evolution, of the feminine, is the life which Giraudoux's Suzanne leads on her tropical island and which I evoked in Chapter 5. Totally free of definitions and models, habits and symbols, she flows into the life which surrounds her, destabilizes herself continuously and lets herself be carried upward by the stream of evolution which has grasped her. Her orientation is not fixed on any place on earth, but on a *telos*, on light—"I am still the only creature which sees the sun in her dreams."

The feminine is at the innermost core of evolution, as the masculine is at the core of planning. Design for evolution must aim at centering in the presence of both. Any single-sided view from the masculine rapes evolution—like Apollo, when he caught up with Daphne and found her turned into a tree, a quivering vegetable, planning often becomes totally ineffective when it tries to force evolution. Apollonic clarity and unambiguity impart power to man; but this power turns against him, if he does not have the wisdom to listen to evolution. And wisdom is bisexual. Teiresias, the figure of wisdom in Greek tragedy, who always puts life into the perspective of temporality and death, is bisexual. Access to this wisdom may be gained through self-abandonment in Dionysian ecstasis, in fluid change. God Dionysus is himself bisexual. He is the mad god not in the sense of chaos but in the sense of a backward link to the origins and the return with a different conception of reality, as in schizophrenia (which has come to be recognized as one of the deepest sources for creativity).

Ecstasis provides the much-needed *re-ligio* backward which ensures fluidity, or at least flexibility, breaks up a solid reality, introduces instability, and thus the possibility for evolutionary mutation, and provides a sense of direction. The best current experiments in "ecstatic" theater—some of which I have enumerated in Chapter 9—provide an idea of what design may mean in a more general sense. In this type of theater, there is hardly the fixed framework of a plot, nor are there characters in the traditional sense. What is shown is a sequence of events which seems to realize its own innate form in dialogue with a strange reality on and off stage. The extremely slow movements of the actors in Japanese Noh or in performances of the Bread and Puppet Theater magnify a thousand-fold the intensity of the most inconspicuous event—raising a glass and drinking, or turning the head, or expressing an emotion by a simple pure gesture of a hand, or the head—letting it flow out and virtually submerge the audience; but it is at the same time acted out with the highest precision.

Design is not a matter for cool heads. Nor is it a matter of getting drunk or going insane. It is inherently bisexual, depending on an intimate blending of freedom and discipline, of disruption and caring. We may call design *Dionysian planning*—clarity abandoning itself to ecstasis in change. Design, in the end, is a religious attitude and experience. It is the marriage of Krishna, or clarity, to Radha, the life force in Hindu mythology. It is the experience of Shiva's dance uniting heaven and earth, making the world pulsate with the rhythm of the universe. It is static duality dissolving into evolution.

The stage for true design is currently set by a growing religious attitude among the young, by an urge toward the collective, and by the return of bisexuality which may be expected to ultimately overthrow two thousand years of cultural bias in favor of male supremacy and misogyny (Hillman, 1972) which were consequently in favor of intellectual ego trips.[1] These are all elements favorable for design, as perhaps the characteristics of any transition or mutatory period are. In times of a well-established and widely shared cultural framework, masculine planning can be effective—at least to the extent that it is aware of the boundary conditions. Planning was

---

1. In this context, it is interesting to note Thom's (1972) belief that sexuality is just the sleep of morphogenesis, a deep breath to prepare the next evolutionary mutation—and thus constitutes a phase characteristic for the period of horizontal stabilization, of the supremacy of the masculine. The feminine is, in fact, much more prone to include bisexuality, than the masculine.

also good enough as long as social space was not so tightly filled by the complexities of highly technological human systems, and as long as the imposed order of man could be upheld. On the other hand, love alone would only suffice if each of us could freely live on a Pacific island such as Suzanne's, not in our complex world with its dynamics and inertias.

Actually, the complexity of our technological world is not a disadvantage, viewed from an evolutionary angle. It means that we will not have to worry about instabilities to initiate a mutation to a higher state—in fact, these instabilities are already making themselves felt in terms of a mounting, multifaceted world crisis with regulation breaking down in many areas, ecological as well as economic, political as well as appreciative areas (Vickers, 1970). The response of planning, of the masculine ego trip, is stabilization at the current level of evolution—stopping evolution; this is the unforgivable error of current prescriptions for curing the crisis (Forrester, 1971; Meadows et al., 1972). But we do not have to fear that such a prescription will stand any chance to be carried out. The objective will has shaken up the world too much already.

Not all of mankind may make the mutation to a higher level. It may well be that a subphylum—the most grossly material and technological subphylum of the Western world, weighed down and made immovable by its own constructions and debris—will phase out in the transition period.[1] Certainly the world will then bemoan a tragic catastrophe for mankind, but evolution is not emotional; nor is that kind of love which I have identified with the objective will. It is at work already in many ways, even in the midst of the Western world. It has produced a spiritual renaissance, especially among the young people in America, a growing awareness of the internal factors of evolution in us, in our bodies, a rapidly deepening understanding of evolution in a scientific way as well as by ways of self-attunement and revelation, a growing wave of mysticism. The instabilities which have set in may still rock us through more highs and lows before the mutation actually occurs, but a major step in evolution, an uplift of mankind, is heralded by them.

The highest-amplitude oscillations, and the evolutionary upward

1. Science fiction is fascinated by an alternative which seems to express one of the oldest archtetypes of humanity, and thus may well contain some evolutionary wisdom: that mankind will leave the earth and "fertilize" the universe with its ideas—a truly evolutionary thought, but perhaps premature for the grossly material version of technological space travel.

swing itself, may be of a nature comparable to turbulent flow in physics. The stabilized state would then correspond to laminar flow. In the laminar state, man may seize control and impose his own determinism. In full turbulence, evolutionary forces may block man from exerting an effective influence; in such phases, human systems may do best to "curl up" and weather it out (Emery, 1967), if they are worth preserving at all. But it is in the phases in which instabilities have built up to considerable size, but do not yet cause full turbulence—we are currently living such a phase—where design has an important role to play. And not only in the build-up phase; design might conceivably also be used to forestall any future stabilization of the kind we had in the past. It might keep human systems much more flexible than heretofore by regulating the processes which are forming and continuously redefining them in a much freer way, encouraging dissent and innovation. By loosening the reins, we will ultimately have more effective control through design than through the traditional alternation between rigidly linear courses and unpredictable massive cultural change.

The big questions which I have raised in the first chapters of this book—What is the role of man in involution, and what will change when social space will have been filled soon?—now seem to find a tentative answer. Man's role will be that of design, of carefully centering his action between subjective and objective will, planning and love. It will no longer be the building of a specific and fairly rigid cultural framework, in the same ways as people build and periodically rebuild their cities and villages within the reach of active volcanoes, or in areas of periodic devastating earthquakes. Rather, it will be man's role to design his world in consonance with evolution, to regulate it in such a way that the cultural mutations become conscious and a matter for design, too, and the whole process of stepping up becomes smoother. In other words, I believe that the phase of evolution on the earth which Teilhard de Chardin calls noogenesis, includes an *active role of man in designing and furthering evolution*—in becoming, individually and through ever more complex human systems, the spearhead of evolution of this planet. In this way, a phylum of humanity may be actively guided through successive evolutionary steps, with human consciousness acting out internal factors of evolution—we have to try, without having the assurance of success. Another phylum may eventually continue where we fail. But certainly, we have the consciousness and the tools to try—and a sense of meaning and purpose of human life which is perhaps the greatest uplift which the idea of evolution has given us. We are neither the manifestations of random fluctuations, nor the dumb children of some unpredictable god; we can indeed shape our own future and the course of evolution—if we flow with the stream, if we *become* the stream.

Design, as I have said, is the interplay of planning and love, of fixing and flowing. Without love, we would lock ourselves into a solid cage (probably a Skinner-box); without planning, we cannot build a human world and play our proper active role in evolution. Without either, humanity would come to an end. Design implies centering and graceful motion; acting out and listening at the same time; letting subjective judgment and objective insight, subjective and objective will, flow together. Design is acting simultaneously at all levels from the micro- to the macroworld, from replacement to regulation, from the material to the spiritual. It is fluidity paired with precision—it is the recognition of *imperfection* as the innermost essence, the energizing principle, of human life. The wisdom of Chuang Tzu, the Chinese poet who wrote more than twenty-two hundred years ago, has cast this idea in a flow of suggestive images which hold much of what I believe the notion of design for evolution to be about:

> He who forces his way to power
> Against the grain
> Is called tyrant and usurper.
> He who moves with the stream of events
> Is called a wise statesman.
> Kui, the one-legged dragon,
> Is jealous of the centipede.
> The centipede is jealous of the snake.
> The snake is jealous of the wind.
> The wind is jealous of the eye.
> The eye is jealous of the mind.
> Kui said to the centipede:
> "I manage my one leg with difficulty:
> How can you manage a hundred?"
> The centipede replied:
> "I do not manage them.
> They land all over the place
> Like drops of spit."
> The centipede said to the snake:
> "With all my feet, I cannot move as fast
> As you do with no feet at all.

How is this done?"
The snake replied:
"I have natural glide
That can't be changed. What do I need
With feet?"
The snake spoke to the wind:
"I ripple my backbone and move along
In a bodily way. You, without bones,
Without muscles, without method,
Blow from the North Sea to the Southern Ocean.
How do you get there
With nothing?"
The wind replied:
"True, I rise up in the North Sea
And take myself without obstacle to the Southern Ocean.
But every eye that remarks me,
Every wing that uses me,
Is superior to me, even though
I can uproot the biggest trees, or overturn
Big buildings.
The true conqueror is he
Who is not conquered
By the multitude of the small.
The mind is this conqueror—
But only the mind
Of the wise man."

LITERATURE REFERENCES

(only sources which are quoted in the book are listed here)

Alexander, Christopher (1967). *Notes on the Synthesis of Form*. Cambridge, Mass.: Harvard University Press.

Assagioli, Roberto (1971). *Psychosynthesis*. New York: Viking Compass Book.

Bailey, Alice A. (1934). *A Treatise on White Magic*. New York and London: Lucis Press.

Beer, Stafford (1971). "The Liberty Machine." In: *Futures*, Vol. 3, No. 4.

Berger, Peter L. (1963). *Invitation to Sociology: A Humanistic Perspective*. Garden City, N.Y.: Doubleday. Also Pelican Paperback (1966), Harmondsworth, Middlesex: Penguin.

Bertalanffy, Ludwig von (1967). *Robots, Men and Minds: Psychology in the Modern World*. New York: Braziller.

—— (1968). *General System Theory: Foundations, Development, Applications*. New York: Braziller.

Böhler, Eugen (1972). "Psychological Prerequisites of Forecasting and Planning." In: *Technological Forecasting and Social Change*, Vol. 4, No. 2.

—— (1973). *Psychologie des Zeitgeistes*. Bern and Frankfurt/Main: Lang.

Bullock, Wynn (1971). *Bullock*. San Francisco: Scrimshaw Press.

Burkhardt, Hans (1973). *Die unverstandene Sinnlichkeit*. Wiesbaden: Limes-Verlag.

Calhoun, John (1970). In: Attinger, Ernest O., ed. *Global System Dynamics*. Basel: S. Karger.

—— (1971). "Control of Population: Numbers." In: Albertson, Peter and Bennett, Margery, eds. "Environment and Society in Transition." New York: *Annals of The New York Academy of Sciences*, Vol. 184.

Campbell, Joseph (1972). *Myths to Live By*. New York: Viking Press.

Castaneda, Carlos (1969). *The Teachings of Don Juan: A Yaqui Way of Knowledge*. New York: Ballantine Books.

——— (1971). *A Separate Reality: Further Conversations with Don Juan.* New York: Simon and Schuster. Also New York: Pocket Books (1972).

——— (1972). *Journey to Ixtlan: The Lessons of Don Juan.* New York: Simon and Schuster.

Churchman, C. West (1968 a). *Challenge to Reason.* New York: McGraw-Hill.

——— (1968 b) *The Systems Approach.* New York: Delacorte. Also Dell Paperbacks.

——— (1972). *The Design of Inquiring Systems.* New York: Basic Books.

Cleveland, Harlan (1972). *The Future Executive.* New York: Harper and Row.

Dorfles, Gillo (1972). "The Pro-Airetic Factor in the Concept of Asymmetry in Design." Paper presented to the conference on "The Universitas Project," Museum of Modern Art, New York, January 1972. Publication of conference proceedings by the Museum of Modern Art forthcoming.

Dubos, René (1969). "Future-Oriented Science." In: Jantsch, Erich, ed., *Perspectives of Planning.* Paris: OECD.

Dürckheim, Karlfried Graf von (1956). *Hara: Die Erdmitte des Menschen.* München: Barth. English translation: *Hara: The Vital Centre of Man.* London: Allen and Unwin (1962).

Eccles, John C. (1970). *Facing Reality: Philosophical Adventures by a Brain Scientist.* New York, Heidelberg and Berlin: Springer.

——— (1973). "Cerebral Activity and the Freedom of the Will." Paper presented at the Second International Conference on the Unity of the Sciences, Tokyo, November 1973. Publication planned.

Eliade, Mircea (1962). *The Furnace and the Crucible: A Study on the Origins of Alchemy.* New York: Harper Torchbooks.

Emery, Fred E. (1967). "The Next Thirty Years: Concepts, Methods and Anticipations." In: *Human Relations,* Vol. 20.

Feyerabend, Paul K. (1967). "Theater als Ideologiekritik: Bemerkungen zu Ionesco." In: *Die Philosophie und die Wissenschaften: Simon Moser zum 65. Geburtstag.* Meisenheim am Glan: Anton Hain.

Fischer, Roland (1971). "A Cartography of the Ecstatic and Meditative States." In: *Science,* Vol. 174, No. 4012.

Forrester, Jay W. (1971). *World Dynamics.* Cambridge, Mass.: Wright-Allen.

Frankl, Viktor E. (1949). *Der unbewusste Gott.* Wien: Amandus.

Freeman, Christopher et al., eds. (1973). *Thinking about the Future.* London: Chatto and Windus.

Fromm, Erich (1956). *The Art of Loving.* New York: Harper and Row.

Furtwängler, Wilhelm (1954). *Ton und Wort.* Wiesbaden: Brockhaus.

——— (1955). *Der Musiker und sein Publikum.* Zürich: Atlantis.

Gabor, Dennis (1963). *Inventing the Future.* London: Secker and War-burg. Also Harmondsworth, Middlesex: Penguin Paperback (1964); and New York: Knopf (1964).

—— (1970). *Innovations: Scientific, Technological and Social.* Oxford: Oxford University Press.

—— (1972). *The Mature Society.* London: Secker and Warburg.

Galbraith, John K. (1967). *The New Industrial State.* Boston: Houghton-Mifflin.

Galtung, Johan (1970). "On the Future of Human Society." In: *Futures,* Vol. 2, No. 2.

Giraudoux, Jean (1921). *Suzanne et le Pacifique.* Paris: Bernard Grasset. English translation: *Suzanne and the Pacific.* London: Putnam (1923).

Glansdorff, P. and Prigogine, Ilya (1971). *Thermodynamic Theory of Structure, Stability, and Fluctuations.* New York: Wiley-Interscience.

Gosztonyi, Alexander (1968). *Der Mensch und die Evolution.* München: Beck.

—— (1972). *Grundlagen der Erkenntnis.* München: Beck.

Habermas, Jürgen (1966). "Verwissenschaftlichte Politik in demo-kratischer Gesellschaft." In: Krauch, Helmut et al., eds., *Forschungs-planung.* München: Hanser.

Hauser, Jürg A. (1973). "Hintergründe der Entwicklungsproblematik." In: *Neue Zürcher Zeitung.* Foreign edition, No. 350, 23 December 1973.

Hillman, James (1972). *The Myth of Analysis.* Evanston, Ill.: North-western University Press.

Holling, C.S. (1973). "Resilience and Stability of Ecological Systems." In: Richard F. Johnston, Peter W. Frank, and Charles D. Michener, eds., *Annual Review of Ecology and Systematics,* Vol. 4. Palo Alto, Cali-fornia: Annual Reviews.

—— and Goldberg, M.A. (1971). "Ecology and Planning." In: *Journal of the American Institute of Planners,* Vol. 37, No. 4.

Huxley, Aldous (1950). *The Perennial Philosophy.* London: Chatto and Windus.

—— (1954). *The Doors of Perception.* New York: Harper and Row.

The I Ching, or Book of Changes (1967). The Richard Wilhelm trans-lation, rendered into English by Cary F. Baynes. Bollingen Series 19. Third edition. Princeton, N.J.: Princeton University Press.

Illich, Ivan (1970 a). "The Dawn of Epimethean Man." Paper contributed to the Conference on "Technology: Social Goals and Cultural Options," Aspen, Colorado, August/September 1970.

—— (1970 b). "Schooling: The Ritual of Progress." In: *The New York Review,* 3 December 1970.

Jantsch, Erich (1967). *Technological Forecasting in Perspective.* Paris: OECD.

——, ed. (1969). *Perspectives of Planning.* Paris: OECD.

―――― (1972 a). *Technological Planning and Social Futures.* London: Cassell/ABP; and New York: Halsted Press.

―――― (1972 b). "Forecasting and the Systems Approach: A Critical Survey." In: *Policy Sciences,* Vol. 2, No. 4.

Jonas, Hans (1969). *The Phenomenon of Life.* New York: Delta Books.

Jung, Carl Gustav (1961). *Memories, Dreams, Reflections.* New York: Vintage Books.

Keleman, Stanley (1971). *Sexuality, Self and Survival.* San Franciso: Lodestar Press.

―――― (1974). *The Body is alive and more.* New York: Simon and Schuster.

Kluckhohn, Clyde (1951). "Values and Value Orientations in the Theory of Action." In: Parsons, Talcott and Shils, eds. *Toward a General Theory of Action.* Cambridge, Mass.: Harvard University Press.

Krippner, Stanley, and Rubin, Daniel, eds. (1973). *Galaxies of Life: The Human Aura in Acupuncture and Kirlian Photography.* New York: Gordon and Breach. Also under the new title *The Kirlian Aura: Photographing the Galaxies of Life,* as paperback, Garden City, N.Y.: Anchor Books.

Kuhn, Thomas S. (1962). *The Structure of Scientific Revolutions.* Chicago: University of Chicago Press.

The Kybalion: A Study of the Hermetic Philosophy of Ancient Egypt and Greece, by Three Initiates (1922). Chicago, Ill.: The Yogi Publication Society.

Laszlo, Ervin (1972 a). *Introduction to Systems Philosophy: Toward a New Paradigm of Contemporary Thought.* New York: Gordon and Breach. Also New York: Harper Torchbooks.

―――― (1972 b). *The Systems View of the World.* New York: Braziller.

――――, ed. (1972 c). *The Relevance of General Systems Theory.* New York: Braziller.

―――― (1973 a). "The Ideal Scientific Theory: A Thought Experiment." In: *Philosophy of Science,* Vol. 40, No. 1.

――――, ed. (1973 b). *The World System: Models, Norms, Applications.* New York: Braziller.

LeShan, Lawrence (1969). *Towards a General Theory of the Paranormal: A Report of Work in Progress.* New York: Parapsychology Foundation.

Levins, Richard (1973). "The Limits of Complexity." In: Pattee, Howard H., ed. *Hierarchy Theory.* New York: Braziller.

Lilly, John C. (1972). *The Center of the Cyclone.* New York: Julian Press.

Marcuse, Herbert (1955). *Eros and Civilization: A Philosophical Inquiry into Freud.* Boston: Beacon Press.

Marney, Milton and Smith, Nicholas M. (1971). "Institutional Adaptation, Part II: Interdisciplinary Synthesis." McLean, Virginia: Research

Analysis Corporation. Extracts have been published under the title "Interdisciplinary Synthesis" in: *Policy Sciences*, Vol. 3 (1972).

May, Rollo (1969). *Love and Will*. New York: W. W. Norton. Also Dell Laurel Paperback (1974).

McKenna, Terence and McKenna, Dennis (1972). *Shamanic Investigations*. Berkeley: Manuscript (publication scheduled for 1975 by Continuum Books, New York: probable title, *The Invisible Landscape*.

Mead, Margaret (1973). "Models and Systems Analyses as Metacommunication." In: Laszlo, Ervin, ed. *The World System*. New York: Braziller.

Meadows, Donella H., Meadows, Dennis L., Randers, Jörgen and Behrens III, William W. (1972). *The Limits to Growth*. New York: Universe Books.

Mesarović, Mihajlo D., Macko, D., and Takahara, Y. (1970). *Theory of Hierarchical, Multilevel Systems*. New York: Academic Press.

—— and Pestel, Eduard C. (1972). "A Goal-Seeking and Regionalized Model for Analysis of Critical World Relationships—the Conceptual Foundation." In: *Kybernetes*, Vol. 1.

Metzner, Ralph (1971). *Maps of Consciousness*. New York: Collier; and London: Collier-Macmillan.

Michael, Donald N. (1968). *The Unprepared Society: Planning for a Precarious Future*. New York: Basic Books.

—— (1973). *On Learning to Plan—and Planning to Learn: The Social Psychology of Changing Toward Future-Responsive Societal Learning*. San Francisco: Jossey-Bass.

Monod, Jacques (1970). *Le Hazard et la Necessité*. Paris: Editions du Seuil.

Morgenstern, Oskar (1972 a). "Descriptive, Predictive and Normative Theory." In: *Kyklos*, Vol. 25, Fasc. 4.

—— (1972 b), "Thirteen Critical Points in Contemporary Economic Theory: An Interpretation." In: *Journal of Economic Literature*, Vol. 5.

Myrdal, Gunnar (1958). *Value in Social Theory*. London: Routledge & Kegan Paul.

Naranjo, Claudio (1972). *The One Quest*. New York: Viking Press.

—— and Ornstein, Robert E. (1971). *On the Psychology of Meditation*. New York: Viking Press.

Neumann, John von (1963). Collected Works, Vol. 6, *The Role of Mathematics*. Oxford: Pergamon Press.

Ozbekhan, Hasan (1969). "Toward a General Theory of Planning." In: Jantsch, Erich, ed. *Perspectives of Planning*. Paris: OECD.

Pattee, Howard H., ed. (1973). *Hierarchy Theory: The Challenge of Complex Systems*. New York: Braziller.

Polanyi, Michael (1966). *The Tacit Dimension*. Garden City, N.Y.: Doubleday.

Popper, Karl R. (1972). *Objective Knowledge.* Oxford: Clarendon Press.

Presman, A.S. (1970). *Electromagnetic Fields and Life.* New York and London: Plenum Press.

Pribram, Karl H. (1969). "The Neurophysiology of Remembering." In: *Scientific American,* Vol. 200, January 1969.

Prigogine, Ilya (1972). "La Naissance du Temps." In: *Bulletin de l'Academie Royale de Belgique, Classe des Sciences,* Vol. 58.

——— (1973 a). "Time, Irreversibility and Structure." In: Mehra, Jagdish, ed. *The Physicist's Conception of Nature.* Dordrecht and Boston: D. Reidel.

——— (1973 b). "Irreversibility as a Symmetry Breaking Process." In: *Nature* (1973).

——— (1974a). "Physique et Metaphysique." In: *Bulletin de l'Academie Royale de Belgique, Classe des Sciences,* Volume Speciale pour le Bicentennaire.

——— (1974b). *Stability, Fluctuations, and Complexity.* Brussels: Manuscript.

——— and Nicolis, Gregoire (1971). "Biological Order, Structure and Instabilities." In: *Quarterly Review of Biophysics,* Vol. 4, Nos. 2 and 3.

———, Nicolis, Gregoire, and Babloyantz, Agnès (1972). "Thermodynamics of Evolution." In: *Physics Today,* Vol. 25, Nos. 11 and 12.

Ram Dass, Baba, vulgo Alpert, Richard (1971). *Remember—Be Here Now.* San Cristobal, New Mexico: Lama Foundation.

Rapoport, Anatol (1970). "Methodology in the Physical, Biological and Social Sciences." In: Attinger, Ernest O., ed. *Global Systems Dynamics.* Basel: S. Karger.

Reich, Charles (1970). *The Greening of America.* New York: Random House.

Roberts, Jane (1970). *The Seth Material.* Englewood Cliffs, N.J.: Prentice Hall.

Rock, Pennell (1973). *The Return of the Person.* New York: Arica Institute.

Scholem, Gershom (1955). *Major Trends in Jewish Mysticism.* New York: Schocken Books.

Sibatani, Atuhiro (1973). *Antiscience: Toward one Knowledge, one Learning* (in Japanese). Tokyo: Misuzu Syoboo.

Siu, R.H.G. (1957). *The Tao of Science: An Essay on Western Knowledge and Eastern Wisdom.* Cambridge, Mass.: M.I.T. Press. Also M.I.T. Press Paperback (1964).

Sutherland, John W. (1973). *A General Systems Philosophy for the Social and Behavioral Sciences.* New York: Braziller.

Taylor, Alastair M. (1973). "Some Political Implications of the Forrester

Model." In: Laszlo, Ervin, ed. *The World System*. New York: Braziller.

Teilhard de Chardin, Pierre (1959). *The Phenomenon of Man*. New York: Harper and Row.

—— (1964). *The Future of Man*. London: Collins.

Thayer, Lee (1972). "Communication Systems." In: Laszlo, Ervin, ed. *The Relevance of General Systems Theory*. New York: Braziller.

Thom, René (1972). *Stabilité Structurelle et Morphogenèse*. Reading, Mass.: Benjamin.

Tobias, Cornelius A. (1973). "Human and Scientific Concepts of Time." Lecture presented in the Symposium "Science and Humanism: Partners in Human Progress," University of California, Berkeley, March 1973. Publication of symposium papers forthcoming, University of California Press.

Toynbee, Arnold J. (1972). *A Study of History*, revised and abridged edition. London: O.U.P.

Tryon, Edward P. (1973). "Is the Universe a Vacuum Fluctuation?" In: *Nature*, Vol. 246, p. 396.

Vickers, Geoffrey (1968). *Value Systems and Social Process*. London: Tavistock Publications; and New York: Basic Books. Also Pelican Paperback (1970); Harmondsworth, Middlesex: Penguin.

—— (1969). "The Tacit Norm." Paper presented at the Burg Wartenstein Symposium No. 44 on "The Moral and Esthetic Structure of Human Adaptation," Wenner-Gren Foundation for Anthropological Research, July 1969.

—— (1970). *Freedom in a Rocking Boat: Changing Values in an Unstable Society*. London: Allen Lane, The Penguin Press; and New York: Basic Books. Also as Pelican Paperback; Harmondsworth, Middlesex: Penguin.

—— (1973 a). *Making Institutions Work*. London: ABP; New York: Halsted Press.

—— (1973 b). "Motivation Theory—a Cybernetic Contribution." In: *Behavioral Science*, Vol. 18, No. 4.

Waddington, Conrad H., ed. (1970). *Towards a Theoretical Biology*. Chicago: Aldine.

Whyte, Lancelot L. (1965). *Internal Factors in Evolution*. New York: Braziller.

Wigner, Eugene P. (1967). *Symmetries and Reflections*. Bloomington, Ind.: Indiana University Press.

Wright, Hendrick von (1963). *Logic of Preference*. Edinburgh: Edinburgh University Press.

Zwicky, Fritz (1969). *Discovery, Invention, Research*. New York: Macmillan.

Name Index

Subject Index